Topics in non-ferrous extractive metallurgy

Critical Reports on Applied Chemistry Volume 1

Topics in non-ferrous extractive metallurgy

Edited by A. R. Burkin

A Halsted Press Book

John Wiley & Sons
New York — Toronto

© 1980 by
Blackwell Scientific Publications

All rights reserved. No part of this
publication may be reproduced, stored
in a retrieval system, or transmitted,
in any form or by any means,
electronic, mechanical, photocopying,
recording or otherwise without the
prior permission of the copyright owner.

First published 1980

Published in the U.S.A. and Canada
by Halsted Press,
a Division of John Wiley & Sons, Inc.,
New York

Library of Congress
Cataloging in Publication Data

Topics in non-ferrous extractive
metallurgy
(Critical Report on applied chemistry;
v. 1)
"A Halsted Press Book"
Includes index
1. Nonferrous metals—Metallurgy.
A. R. Burkin, Alfred Richard.
II Series
TN758.T65 669   80-17435
ISBN 0-470-27016-0

Printed in Great Britain

Contents

Editor's Introduction

Western civilisation depends on three great primary industries, production of food, production and utilisation of minerals, and the provision of energy. These are closely interdependent and each is said to be moving into a stage of being unable to satisfy world demand.

The reviews contained in this volume are concerned with methods of producing certain metals from mineral resources. It is fitting that they should be published under the auspices of the Society of Chemical Industry because they show the great efforts which have been made in recent years to improve existing processes and to devise new ones which, in many cases, involve chemical reactions which have never before been used in extractive metallurgy.

There are several reasons why changes in the technology are necessary. The first is that growing restrictions on the discharge of gaseous, liquid and solid wastes from metallurgical plants make them essential. The second is that it is becoming necessary to use ores which cannot be economically treated by processes used conventionally at present, and a third is that a new process or change in an existing one may improve efficiency under some, if not all, circumstances.

It is frequently stated that the world's reserves of ores of some metals are becoming exhausted. Such an assertion in a particular case requires a large amount of factual and technical justification before it can be accepted. An attempt has been made to show what happens when supply and demand for a metal are seriously different, in the review dealing with uranium.

In some cases exhaustion of reserves is equated to diminishing reserves of high grade sulphide ores which can be treated in modern smelters. Copper smelting originally used hand-picked lumps of ore. As this became impracticable the froth flotation process was introduced early this century, producing high-grade concentrates on which modern copper smelting practice has been developed. As the grade of sulphide ore available in a mine becomes lower, the cost of mining and milling it becomes higher per unit of copper contained in the concentrates, until the economic cut-off grade is reached, when production stops. The copper sulphides remaining in the ore-body are then no longer an actual reserve of copper ore unless either the real price of copper rises or a new or improved process is introduced which makes it possible to recover the metal from it economically. This provides the incentive to work out methods for treating low grade sulphide ores.

A different requirement is to develop processes which use the so-called oxide ores. Thus most of the known reserves of nickel occur in the form of silicates, the laterite ores, whereas most of the nickel produced today is obtained from sulphide ores. A change to laterites must come in due course and the major nickel producing companies have been developing new or updated processes for treating these ores. The recovery of metal values from deep sea nodules would be another example of treating a new source, should they ever be 'mined' on a significant scale.

Some copper sulphide ore bodies have been subjected to weathering at the top due to reaction with water containing dissolved oxygen. Over geological periods of time this caused the formation of an 'oxide cap', containing copper in various minerals other than sulphides, as well as the distribution of copper over areas around the ore body due to reprecipitation from ground water. Only limited amounts of oxide ore can be fed into a smelter together with the sulphide flotation concentrates, so that most of it was an unused potential source of copper.

The solvent extraction – electrowinning process which was proved to be commercially successful in the late 1960's is now being used to recover copper from these oxide ores. In this process an organic chelating ligand, of a type described below, is used to selectively extract copper from an impure acidic aqueous sulphate solution, into a hydrocarbon diluent. The copper is stripped from the organic phase by a more strongly acidic solution of copper sulphate coming from the electrolytic tank house, to which it returns with its copper content increased. A number of plants were built to use this process, one in Zambia to treat 14,000 gallons per minute of impure copper-containing solution. Others were planned but the fall in the price of copper caused postponement of the introduction of more production capacity.

Any kind of ore containing copper can be used to feed the solvent extraction – electrowinning process provided that its copper content can be dissolved so as to produce an acidic copper sulphate solution. This includes low-grade sulphide ores from which concentrates cannot be made economically, and waste rock from the mining operations. The heaps of material are terraced and sprayed with the leaching solution which is collected at the bottom and treated for copper removal before being returned to the heap. The oxidising agent which is required to leach the sulphides is ferric sulphate, which is produced from the rock, and the whole leaching and oxidation sequence of reactions is greatly accelerated by the natural presence of certain bacteria.

The progression from heap and dump leaching has been to leaching the rock *in-situ*, without mining it at all. This has been done in the case of copper ores by fracturing the low-grade ore remaining after mining, using conventional explosives. This technique has also been used on uranium ores and, for these, a later development has been to leach porous sandstone deposits containing uranium without disturbing the ground at all except for drilling a pattern of holes to inject and recover solutions. It seems likely that in a later stage of technological development blocks of ore will be fractured and the surrounding rock used as a container in which pressure leaching will be carried out to recover the metal values.

At the same time as development of the use of low-grade and new types of ores, work on new processes to treat conventional ores and concentrates has been going on. Those concerned with copper sulphide concentrates, dealt with in this volume, would be in direct competition with smelting. It is certain that in a large scale operation smelting will be more economic than any hydrometallurgical process for the treatment of concentrates if considerable quantities of sulphur dioxide may be discharged into the atmosphere. As emission controls become more stringent, and so more expensive to meet, so the difference in cost will lessen. For a small scale operation a hydrometallurgical process might in any case be more economic than smelting.

One of the incentives to develop new processes for treating copper concentrates was the belief that they would be 'pollution-free'. This is not the case, but one advantage of some of them is the fact that in certain of the new processes some at least of the sulphur content of the ore can be recovered in elemental form. It might also be thought that pyrometallurgical process would require the use of more energy because they work at high temperatures. However, most of the heat is obtained by oxidation of the sulphides, whereas the energy required to pump solutions in hydrometallurgical processes is very considerable.

In the cases of many other metals new processes have been introduced into the overall flowsheets. These often involve solvent extraction and this technique is having a very significant effect on non-ferrous extractive metallurgy in general. It was first used in the uranium industry and by the mid 1960s it seemed that it would not extend beyond that. Its use in copper recovery caused it to be looked at again and it was realised that in chloride solutions some metals can be made to form anionic or uncharged species which permits their separation from others which remain as cations in the same solution. Uranium forms an anionic complex in sulphate solutions and it was this fact which led to the use of ion exchange and solvent extraction in its recovery and purification.

Increasing understanding of solvent extraction and the solution chemistry of metals in general, among those working in the field of hydrometallurgy has led to the development of numerous processes which can be fitted into overall flowsheets. These do not seem to be known to anyone except those actively working in the field. It is in the belief that they may be of interest to many other chemists and engineers that the three authors of the reviews contained here have written them.

Jargon

As one of those chemists the Editor of the Society said that it would help if some of the more obscure terms used in the text were explained. Most of them could not be replaced by other words without making the text appear strange to people acquainted with the fields of hydrometallurgy and mineral processing, and it is hoped that some of these will read it. Explanations are given below.

The size of particulate solids is expressed as a range of sieve mesh sizes. These sizes were formerly given as the number of wires to the inch but are now usually expressed in micrometres. Thus $+74\ \mu m$ means that the particles are larger than $74\ \mu m$ or, more accurately because of their irregular shapes, they will not pass through a 200 mesh screen which has holes defined in tables as $74\ \mu m$. $-74\ \mu m$ particles will pass through a 200 mesh screen. Although it may seem odd to speak of particles of size about $74\ \mu m$, the number is given as it appears in the table of screen sizes.

Leaching of particulate ores is frequently carried out in pachuca tanks. These are vertical cylindrical tanks with a conical bottom, air usually being introduced through the bottom point of the cone to cause agitation. A mechanical stirrer may also be used if air alone does not give sufficient mixing to prevent solid separating out. The height of a tank may be 10 metres or more, with a similar diameter.

After leaching, the solution is known as the 'pregnant' liquor. A statement that a particular element 'reports' in the leach residues means that it is found in the residues after leaching.

A number of proprietary extractants for metals are referred to in several places in the text. The chemical nature of the materials is believed to be as follows.

LIX reagents from General Mills Inc (now Henkel Corporation)

LIX 63

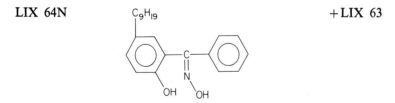

LIX 64N $+$ **LIX 63**

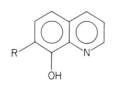

LIX 70   Has Cl adjacent to the OH on the benzene ring above.

KELEX reagents from Ashland Chemical Co

ACORGA reagents from Imperial Chemical Industries Ltd (Acorga Ltd)

P17

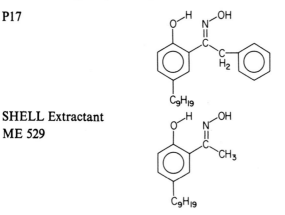

SHELL Extractant
ME 529

Finally, when clarifying all these matters with the Editor, I felt that he looked quite sorry when I wrote 'metallic iron originating in mining and comminution' in place of 'tramp iron'.

14th August 1979   A. R. Burkin

Extractive metallurgy of copper

Gunnar Thorsen

1   Introduction

One could hardly think of producing any review on non-ferrous extractive metallurgy without taking into account the metal copper. Although books, reviews, and special issues on copper have been numerous in recent years[1-7] the ever changing conditions, particularly in the energy and environmental aspects of modern industrial activity – not to mention the political impact – seem to make a venture into the world of copper always worth while. Even for those who follow the copper scene from the ringside there will always be something new to add to the story.

The new matters do not necessarily need to be innovatory in nature, in the strictest sense. Due to the dynamic changes in the field a re-evaluation of earlier

1

findings and previously suggested processes may reveal some approaches which could well turn out to be of importance and interest under present circumstances.

One aspect of the production of copper is concerned with the basic choice between pyrometallurgical and hydrometallurgical routes. The traditional pyrometallurgical production by copper smelters became, more or less overnight, the bad wolf in the environmental and conservation discussions which arose in the sixties. It was true that no breakthrough had been made within the copper smelting industry which could enable the new regulations on sulphur dioxide emissions to be met, at least not without enormous investment which most of the industry itself was not able to make. However, new technology was on its way – the pyrometallurgists were put under pressure and innovations, largely based upon the principles of flash smelting and continuous processing, were launched. Although still in a state of development, the first new technology smelters are now in operation.

In the meantime, the hydrometallurgical approach to copper processing has received a lot of support. From the environmental point of view this appeared to be the answer to avoid pollution problems. However, the hydrometallurgical processes have pollution problems of their own. Problems associated with gas emission may appear to be nonexistent, but plant effluents in the liquid and solid states may lead to serious contamination of the environment if not taken care of properly and quantitatively. Usually there are technical solutions available, but again the investment and operating costs may be considerable and may even prevent an overall profit being made.

This review on the extractive metallurgy of copper is not intended to be comprehensive, attempting to cover all aspects of the activity taking place at the present time in the field of copper extraction. Rather, it will be selective and an attempt has been made to present the main ideas reflected in the various processes actually in operation or being developed.

Obviously, a number of parallel developments are taking place in different plants and laboratories around the world. In the hydrometallurgical processing of copper concentrates, for instance, the chloride leach routes seem to be much favoured. In the pyrometallurgical field, on the other hand, there are several different kinds of approach being made to try to integrate the roast-smelting-converting steps into an autogenous continuous process. The differences in approach appear mostly to be found in the engineering techniques and design of the reactors rather than in the process principles.

The descriptions of processes outlined in this review will for obvious reasons have to be based on published data or other information at hand. From the vast amount of material published during recent years it will always be a matter of opinion and taste as to what to take into consideration. One criterion for selection could be the degree of *simplicity* in the processing circuits. Elegance in extractive metallurgy, as in all industrial processing, appears as short cuts which minimise the route from raw materials to products. In addition the possibility of complete utilization of all components present in the raw materials will always find favour.

It is believed that important key factors in the present and future processes for recovering copper from base materials, i.e. ores and concentrates, will be the handling of the main by-products, iron and sulphur. Special attention will therefore be paid to the ways in which these components are taken care of, either by disposal and rejection or, better, by making useful and saleable products from them.

The basic copper extraction industry has traditionally used two main approaches: pyrometallurgical and hydrometallurgical processing. It is to be noted, however, that the general trend of modern extractive metallurgy tends to integrate these approaches. This is reflected for instance in the roast-leach-electrowin processes for both copper and zinc where in the front-end of the flowsheets the roasting step is a typical pyrometallurgical stage while the leaching of the calcine product and further electrowinning of the metal values belong to hydrometallurgy. At the far end of the flowsheet the techniques of pyrometallurgy may be revived, as for instance in the process for smelting of a lead-silver residue.[8]

In the copper industry, however, a pyrometallurgical process has a distinctive feature and meaning in that the copper is produced in a furnace as liquid metal (blister copper). Basically in this review that outline and definition will be followed.

2   Pyrometallurgical processing of copper sulphide concentrates

About 90% of primary copper production is from sulphide ores, which contain from 0·5 to 2% Cu in the orebody. After concentration by froth flotation the resulting copper concentrate will contain from 20 to 30% Cu which will be the typical starting point for a pyrometallurgical smelting process.

The pyrometallurgical process may be divided into the following three steps:

1   *Roasting.* Sulphur will be partially eliminated as SO_2 by partially oxidising the sulphides at 500–700°C by reacting with air.

2   *Matte smelting.* The roasting product is melted together with a silica flux at about 1200°C, forming a liquid two-phase system consisting of a sulphide phase (matte), containing virtually all the copper, and a liquid slag phase containing silica, alumina, iron oxides and other minor oxides with as low copper content as possible.

3   *Converting.* The liquid matte containing copper and iron sulphides is oxidised with air whereby the converting reactions take place in two sequential stages:

(i) the FeS elimination stage, i.e.

$$2FeS + 3O_2 + SiO_2(flux) \rightarrow 2FeO.SiO_2(slag) + 2SO_2$$

(ii) the blister copper formation stage, i.e.

$$Cu_2S + O_2 \rightarrow 2Cu + SO_2$$

An essential feature of the converting reactions is that the second reaction – the blister copper formation – does not start before there is only about 1% FeS left in the matte.[1]

It will be seen that steps 1 and 3 release sulphur dioxide. Both steps are controlled oxidation processes, however, and in modern copper smelting practice the demand for efficient SO_2 recovery has to be adequately met. The concentration of SO_2 in the effluent roaster gases is high enough (5–15%) for it to be efficiently used for production of sulphuric acid, or even reduced to elemental sulphur. Furthermore, the oxidation reactions may be carried out using oxygen enrichment. In this case the energy used in heating the nitrogen in the effluent gas when air is used for combustion is saved. The main effect of oxygen enrichment, however, is to increase the SO_2 concentration. In the INCO flash smelting process, which will be discussed later, the content of sulphur dioxide in the effluent gases may be as high as 80% SO_2. Obviously, this is an excellent source for production of liquid SO_2 or elemental sulphur.

2.1   Conventional processes

The conventional practice of most copper smelters has been to carry out the three steps of roasting, smelting, and converting in separate operations. The roasting, which is an optional step, is often practised prior to smelting with the purpose of utilising the released heat to dry and heat the charge before it is added to the smelting furnace. This operation will result in a considerable saving of fuel in the smelting process and also increase the capacity of the smelting furnace.

At the second stage – matte smelting – the real challenges to smelting techniques are the following requirements:
1, the abatement of the SO_2 emission problem; 2, minimising the fuel and energy requirement; 3, rejection of a waste slag phase sufficiently low in copper.

The conventional and widely used practice has been to use a reverberatory furnace to carry out the melting of the mixture of copper concentrate/calcine and the silica flux. The heat for smelting and maintaining the temperature of the two-phase liquid system of matte and slag at about 1200°C is provided by combustion of oil, natural gas or pulverized coal. The hot combustion gases sweep and pass over the charge.

Apart from the considerable consumption of hydrocarbon fuels, the inevitable oxidation of sulphides in the smelting furnace will result in a serious pollution problem because the SO_2 in the smelter effluent gases is too dilute (1–2%) for recovery as sulphuric acid. In modern reverberatory plants which usually are provided with a fluo-solid roaster it is possible to reach 90% sulphur recovery without treatment of exhaust gas from the smelter[6]. Although an overall 10% emission of SO_2 might be tolerated in some remote areas even in the future, it is not believed that any new reverberatory furnaces will ever again be built.[9]

The problem of SO_2 emission can be solved by using an electric furnace for smelting. The heat for melting is generated by the electrical resistance of the slag, passing electric current between carbon electrodes immersed in it. The quantity of effluent gas will in this case be rather small and the SO_2 concentration can readily be controlled by adjusting the air entrainment into the electric furnace.

The electrical energy is used efficiently and the electric furnace is very versatile and flexible for the treatment of all materials. The major disadvantage is that electricity tends to be rather costly, depending on the area of location.

An important point to consider in the matte smelting stage is the copper content of the slag phase. The copper concentration in the discarded slags varies from 0·2 to 1% depending upon the particular type of operation.[1] Minimum solubility of copper is obtained by keeping the silica concentration in the slag close to saturation. This will also keep the matte-slag miscibility at a minimum. Furthermore, the formation of magnetite, Fe_3O_4, should be kept as low as possible by avoiding excessively oxidising conditions. Magnetite will increase the viscosity of the slag, thereby making separation of the two phases more difficult.

The purpose of the converting stage for the copper matte is to remove iron, sulphur and other impurities from matte, thereby producing liquid metallic copper of about 99% purity (blister copper). Converter slags contain from 2 to 15% Cu and have to undergo treatment for copper recovery, usually by froth flotation of the copper from solidified and slowly cooled slag.

2.2   *Modern process techniques of copper smelting*

An overall energy balance for the copper smelting process shows that considerable energy is obtainable by complete oxidation of the sulphide charge. Thus the trends in modern copper smelting are to make full use of this energy source. Efforts are also directed towards making the processes suitable for continuous operation.

The first development along these lines was the introduction of flash furnace smelting which makes considerable use of the sulphide-combustion energy for the melting of matte and slag. Thus the treatment of chalcopyrite by flash smelting can be represented by the equation:

$$2CuFeS_2 + 5/2\,O_2 + SiO_2(flux) \rightarrow$$
$$Cu_2S\,.\,FeS(matte) + FeO\,.\,SiO_2(slag) + 2SO_2 + heat$$

The flash smelting technique has been developed on the basis of two different principles:

1 The INCO process which is made completely autogenous by use of commercial oxygen.

2 The Outokumpu process which uses preheated air, in some cases with oxygen enrichment. Hydrocarbon fuel has to be added in this process to make up for the energy difference.

Biswas and Davenport have presented and discussed the two processes at length[1]. In a comparison of them, they state that the major difference of importance is that the INCO process relies completely upon oxygen to make it autogenous while the Outokumpu process uses oil to make up its thermal deficit. On this basis the INCO process will have several advantages over the Outokumpu process:

- it has a much lower overall energy requirement;
- in the absence of nitrogen and hydrocarbon combustion products the volume of effluent gas is small and consequently the gas-collection equipment is small;
- the SO_2 concentration in the combustion gases of the INCO process will be as high as 80% which obviously makes them well suited for further processing to sulphuric acid, liquid SO_2 or elemental sulphur;
- dust losses will be low due to the modest volumetric flow of gas;
- throughput in the same size of equipment will be about 30% higher in the INCO process than in the Outokumpu process.

From the later developments[10] it has been noted that oxygen enrichment has been introduced in the Outokumpu process. It is further reported that one trend in the metallurgical development of flash smelting has been towards the continuous production of high-grade matte or blister copper. By this means batch converting has been reduced considerably or even eliminated in some cases, depending on the concentrate material. A similar approach of integration of smelting and converting has also been reported recently by INCO[11] when a pilot-plant campaign was undertaken to extend the oxygen flash smelting technique to a smaller smelting unit which would permit smelting and converting in a single unit without fuel requirements.

These later developments in the INCO and Outokumpu flash smelting processes indicate the general modern trend towards a fully integrated and continuous operation of the roasting-smelting-converting steps in pyrometallurgical extraction of copper from sulphide concentrates.

A continuous operation seems particularly attractive from the point of view of eliminating the conventional separate tapping operations in the batch processes. In the approach of combining the roasting, smelting and converting steps into one furnace or reactor, however, obviously the engineering and the design of the reactor will be of the utmost importance. In this context the thermodynamic analysis of the *Q-S process* for coppermaking[12] is very instructive and valuable. It has been shown by this analysis that an efficient operation of the single furnace reactor requires the establishment and control of gradients in oxygen activity by a stagewise introduction of oxygen and solid feed materials along the reactor. A schematic diagram of the staging in the Q-S converter is shown in Fig. 1. This model takes into account the thermodynamic conditions for an optimum quality of the copper metal product. Of particular interest is the thermodynamic discussion of deconverting and slag cleaning in order to produce a discharge slag suitably low in copper (for example 0·2% Cu).

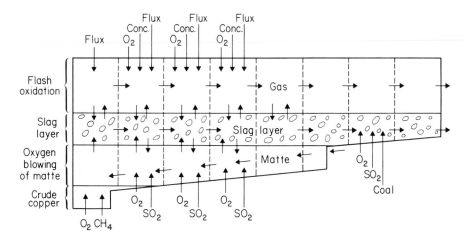

Fig. 1. Schematic diagram of staging in Q–S process[12].

Three continuous processes actually tried in operation are presented in the following.

2.3   The WORCRA process

In continuous production of blister copper from matte in a single furnace, three liquid phases will be present: slag, matte and blister copper. Fig. 2 shows the furnace used in the WORCRA process (named after the inventor H. K. Worner and Conzinc Riotinto of Australia Ltd).

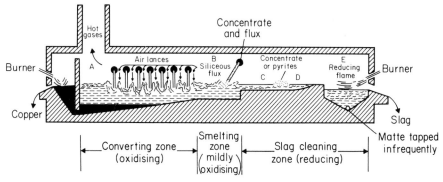

Fig. 2. The WORCRA process pilot reactor[1].

The concentrate and flux are introduced in the mildly oxidising smelting zone. The matte and slag flow countercurrently and in a slag cleaning zone the copper concentration in slag is reduced to about 0·5% Cu. This comparatively low concentration of copper is accomplished by using a rather large part of the furnace for the slag cleaning zone. To keep the temperature at the appropriate level, hydrocarbon fuel is burnt in the settling area. It is to be noted that the copper concentration of the slag may be lowered down to 0·3 % Cu by the addition of pyrites to provide further reducing conditions in the slag cleaning zone.

Although the combustion gases from the slag cleaning contain only 1–2 % SO_2 the high-strength gases obtained from the oxygen enriched oxidising zone make the overall SO_2 concentration suitable for sulphuric acid production.

In spite of the energy requirement the WORCRA process undoubtedly appears to be very promising, particularly because the slag phase may be directly rejected without milling and refloating. However, the process has so far only been operated in a pilot plant which was closed down in 1970.

2.4   The Noranda process

The only one-furnace continuous process for making blister copper actually in industrial operation seems to be the Noranda process. Fig. 3 shows the Noranda reactor which has much in common with an ordinary elongated converter.[1,13] The

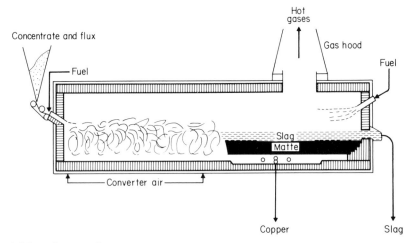

Fig. 3. Noranda reactor[2].

reactor is charged with concentrates and flux at one end where the solids are heated with a burner flame and the smelting takes place when air or oxygen-enriched air is injected into the matte layer. Slag and matte move along the reactor with the air/oxygen blowing through submerged tuyeres, the converting taking place continuously along the reactor. The blister copper is tapped from a 'well' at the bottom and the slag at the far end of the reactor.

The high SO_2 concentration of the gas which is efficiently collected by a hood arrangement is an advantage of the process. However, the simplicity of the one-stage process, being more or less at the condition of equilibrium, has to be paid for by the low quality of the product. The blister copper contains as much as 1–2 %

sulphur which has to be removed in a separate stage. The slag from the reactor contains 8–10% Cu which is recovered by milling and froth flotation operations, the final tailings containing 0·5% Cu. Further development by Kennecott Copper Corporation[14] has cleaned the slags from a Noranda Reactor to 0·2–0·3% Cu in an electrical furnace which employs mechanical mixers to promote magnetite reduction and copper extraction at high rates.

2.5   The Mitsubishi process

The advantage of the counter-current principle of the WORCRA process has been met in the Mitsubishi development which is a true overall continuous process for the production of blister copper from concentrates.[15] The Mitsubishi process operates with three furnaces, one for each of the three metallurgical stages: smelting, converting, and slag cleaning. The furnaces are interconnected as shown in Fig. 4. The melted liquid phases flow by gravity in continuous overflow between the furnaces.

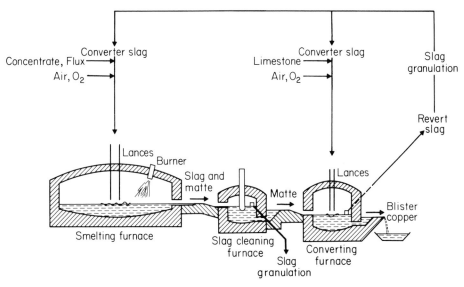

Fig. 4. Schematic view of the Mitsubishi continuous furnace line[16].

This arrangement allows some flexibility in choosing optimum conditions for the oxygen potential which should be higher in the converting stage than in the smelting stage to minimise the copper losses.

To operate the process under optimum conditions, the variables have to be strictly controlled.[8] This is done by an advanced feed-forward automatic control system. Because there is only one reaction zone in each furnace this will allow a high smelting rate. Accordingly there is a high throughput in the system which is of advantage for the overall heat balance.

The rejected slag from the electric slag cleaning furnace will contain 0·5% Cu. The by-product slag from the converting furnace which contains 10–15% Cu is solidified and recycled to the smelting furnace. It is to be noted that no silica is added to the converting furnace which keeps the recycle small. Of major importance, however, is that the silica-free slag in the converter is basic, thereby lowering the sulphur content of the blister copper to 0·1–0·9% S. This is to be compared to the single-stage copper quality of 1–2% S which is obtained in the Noranda process.

2.6   Which process to choose?

This question can never be answered definitely without taking into account a number of local and specific conditions associated with a particular situation. The raw material has to be analysed carefully, but in metallurgical processes unexpected surprises, caused for instance by the behaviour of impurities, may arise. In this context it is reported[17] that the blister copper produced by a continuous one-furnace process tends to contain more antimony, arsenic and bismuth than the blister copper produced by conventional smelting and subsequent converting. The reason for this seems to be that the sulphides of these elements are extracted into the metallic copper always present in the one-stage operation. In the conventional converter the impurity sulphides will be volatilised during the converting before the metallic copper starts to form.

Texasgulf Canada is presently constructing a new copper smelter at Timmins in Northern Ontario which was expected to start up in late 1979. In a very interesting and detailed description of the design development,[16] which includes a thorough examination of alternative processes and plant visits all over the world, the major reasons for selection of the Mitsubishi continuous smelting process are given and may be summarised as follows:

● tried and proven, has operated for some time on a commercial scale;
● not labour intensive, successfully automated with furnace feed, air, oxygen, and fuel inputs being computer controlled;
● molten furnace products merely overflow with the conventional separate tapping operations eliminated;
● capital cost competitive with that for a conventional or flash smelting plant;
● advanced technology not likely to become obsolete in the next 10 or 20 years;
● high quality of blister copper produced and a low-copper slag discarded;
● 99% SO_2 recovered as sulphuric acid, good environmental control and working conditions are expected, few manual metallurgical techniques are required.

In a comparison of continuous copper-making processes Biswas and Davenport[1] state that the difference between the Noranda, Worcra and Mitsubishi processes are due to the method of injecting the air (or oxygen-enriched air) into the matte. After a discussion of the various problems associated with the various methods, their conclusion is that only time will tell which is the best.

There is no doubt that the development within the pyrometallurgical routes for treating copper sulphide concentrates has been brought to a very advanced technological level within the last decade. For a reasonably uniform concentrate of chalcopyrite with a high content of precious metals it seems difficult to envisage a more economic approach on the large scale than a more or less autogenous integrated smelting process. However, special circumstances will be met with all over the world in the extractive metallurgy field. With the advanced level of the modern *chemical* industry, the possibilities of using low-temperature chemical reactions on the commercial scale in the metal industry are increasing.

In the following sections these possibilities are outlined by examining the field of hydrometallurgy, which deals with techniques of producing metals by reactions taking place in liquid aqueous and organic phases.

3   Hydrometallurgical processing routes for copper

For the treatment of complex ores which are difficult to upgrade by physical means there will constantly exist a need for alternative routes. Processing by hydrometallurgical techniques will most likely be the only way to deal with these materials because of the prohibitive energy requirement for heating up large quantities of worthless materials in smelting processes. Furthermore there are many options available in the further treatment of solutions of metal salts. In this context not only aqueous solutions of the metals should be considered but also the use of organic solutions. This brings us to the modern separation technique of solvent extraction, which has introduced a new dimension into the hydrometallurgical processing of metals.

In particular, for extracting copper ions from aqueous leach solutions a number of specific selective organic reagents have been developed and are available from several manufacturers. On the market there are at least four product series: the LIX reagents from Henkel Corporation; the KELEX reagents from Ashland Chemical Co; the Shell Metal Extractant 529 (SME 529); and the Acorga P-5000 Series from Imperial Chemical Industries Ltd (Acorga Ltd).

These extractants are all chelating agents acting as cation exchangers by releasing hydrogen ions while selectively extracting copper ions:

$$2\overline{RH} + Cu^{2+} \rightleftarrows \overline{CuR_2} + 2H^+$$

(A bar over a substance indicates that it is present in the organic phase which comprises the extractant in a diluent, usually a hydrocarbon.)

The LIX and the SME series are derivatives built around hydroxy oxime groups. In the Kelex series the chelating property is based upon 8-hydroxy quinoline while the Acorga P-5000 is developed from the parent salicylaldoxime.[18-22]

In addition to being selective for copper, particularly in preference to iron, the extraction will take place directly from acid leach solutions. Using the normal solvent extraction techniques for copper liquors, the acid produced by the extraction reaction shown above need not be neutralised. The extracting power of the chelating agents with respect to the acidity level of the leach liquor varies and this is reflected in the pH dependence of the equilibrium constants of the extraction reaction.

However, if a strongly extracting reagent is used, the extracting power has to be paid for when the reverse reaction takes place, stripping the copper from the organic phase with sulphuric acid for recovery by electrowinning from an aqueous sulphate solution. Thus for LIX 70 it is reported that at least 300 g/l H_2SO_4 is required for stripping, which would not be suitable for electrowinning.[19] At this high sulphuric acid concentration, however, the recovery of copper as copper sulphate crystals after stripping from the organic phase could be an alternative to electrowinning.

Fig. 5 shows in principle the flowsheet for copper production by leaching, solvent extraction and electrowinning. The ideal process consists of three closed-loop

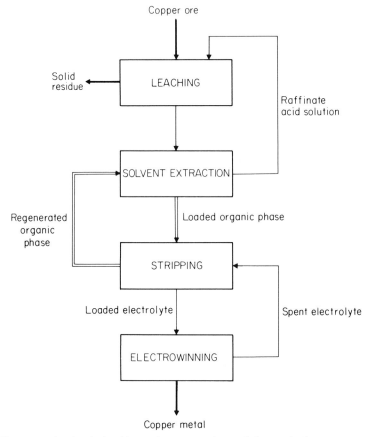

Fig. 5. Copper production by leaching, solvent extraction and electrowinning.

circuits where the solvent-extraction loop will bring the copper selectively from a dilute leach solution (1–5 g/l Cu) into the organic phase from which it will subsequently be stripped into an acid sulphate electrolyte at a high concentration of 30–50 g/l Cu, suitable for electrowinning.

This kind of processing is particularly suited for treating low-grade oxide ores and sulphide mine wastes where the leaching is a low-cost operation such as bacterial leaching and heap, dump or *in situ* leaching with sulphuric acid. These leaching operations are dealt with in the chapter on uranium extraction in this volume.

3.1   Vat leaching and agitation leaching of oxidised copper ores

Vat leaching has usually been carried out in order to produce a pregnant leach solution of sufficient copper concentration (30–50 g/l Cu) for direct electrowinning. The leaching takes place by immersing crushed ore in 50–100 g/l sulphuric acid solutions in large rectangular vats. Sequential leaching in a series of vats with a continuous flow of leach acid through the vat system is the normal operating practice.

With the introduction of the solvent extraction technique there is a trend to operate at a much lower copper concentration in the leach circuit to optimise the overall production line. Furthermore, the modern plant designs seem also to favour the more controllable agitation leaching in tanks in series with the agitation provided by air (pachuca tanks) or by mechanical means.

This is the case in the new Anamax oxide plant at Twin Buttes, Arizona[23] which came on stream in 1975. The main copper mineral in the orebody is chalcopyrite ($CuFeS_2$), but cap oxidation has progressed to depths of several hundred feet with a copper content above 0.6% Cu. From stockpiled sources with a grade of about 1% recoverable copper, there is now production at a capacity of 10,000 tons per day of ore. Fig. 6 shows the flowsheet of the Anamax copper oxide plant.

After grinding and milling, the oxide ore material is leached in five mechanically agitated, rubber-lined leach tanks. After a total residence time of about 5 hours most of the soluble copper has been leached. It is interesting to note that the consumption of sulphuric acid (added as 93.2%) is up to 250 lb per ton of ore. This works out to be more than seven times the equivalent acid required to leach the 1% copper content. This example illustrates quite clearly the importance of local conditions and the integrated operation of two different plants, since it is reported that the acid is supplied in tank cars from a nearby smelter.

The pregnant liquor containing about 2.5 g/l Cu is extracted to 0.08 g/l Cu in four stages with LIX 64N. Stripping is carried out in two stages producing an electrolyte solution of about 50 g/l Cu and 90 g/l H_2SO_4 for electrowinning. The copper drop in the tankhouse is from 50 down to 25 g/l Cu in spent electrolyte.

A similar approach to that at the Anamax oxide plant seems to be contemplated

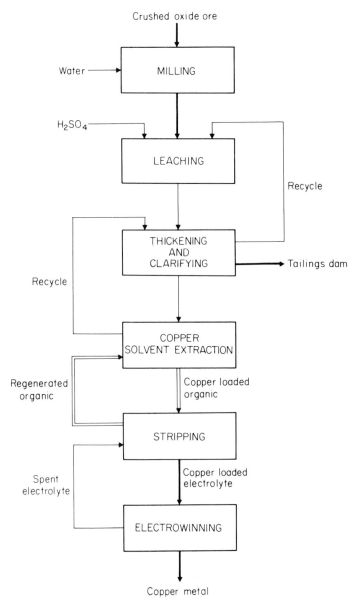

Fig. 6. The ANAMAX oxide plant[23].

for the metallurgical treatment of Chingola cupriferous mica ores in Zambia.[24] The micaceous ores are found in the upper levels of most of the sulphide orebodies in the Copperbelt. The cupriferous mica mineralogy is quite complex, the copper occurring as malachite, pseudo-malachite, crysocolla and cuprite. The copper content usually varies within the range 0·5–1·5%. The ore is readily disintegrated to a very fine size, with comparatively little grinding energy, which is of advantage for the leaching step. From a number of laboratory tests it has been found that due to the slow leaching rates at ambient temperature, a hot dilute acid leaching at 60–70°C will probably be more economic. Furthermore, the use of agitation leaching is favoured because of the control of the leaching reaction.

The proposed flowsheet for the process is shown in Fig. 7. It should be noted that the solvent extraction technique is included, which in this case plays a very important role due to the significant quantities of magnesium, iron and aluminium which will be dissolved from the mica and which make other separation techniques difficult and expensive.

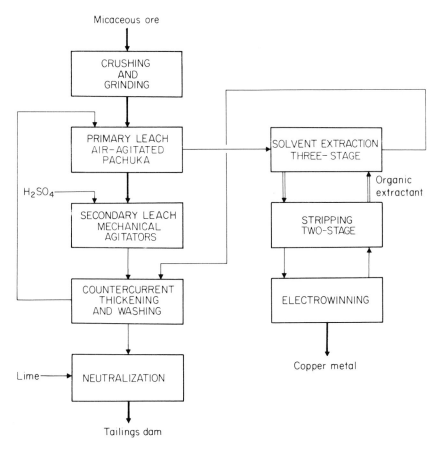

Fig. 7. Proposed treatment route for micaceous ore[24].

It is interesting to note that it is stated that the main drawback of the proposed treatment route for the Chingola micaceous ore is the large quantity of sulphuric acid required. From the test results[24] it may be seen that about 90% of the copper has been leached from the mica sample with an acid consumption of 6 g H_2SO_4 per gramme of Cu. Although this is almost four times the quantity required for the equivalence of copper and sulphate, in the Anamax plant the acid consumption is almost twice as much again per gramme of copper produced.

Again this is confirmation of the importance of local conditions which are shown in the stated conclusion that the required sulphuric acid has to be provided by additions to the smelter gas collection and acid manufacturing facilities of the Zambian Copperbelt.

3.2   Hydrometallurgy for copper concentrates

Of the copper-bearing minerals chalcopyrite ($CuFeS_2$) is by far the most abundant but represents also the main challenge in copper hydrometallurgy due to its comparatively modest reactivity in leaching media. The chemistry of processing chalcopyrite looks attractive, however, in that the principal leach reaction with sulphuric acid may be represented by:

$$CuFeS_2 + H_2SO_4 + \tfrac{5}{4}O_2 \rightarrow CuSO_4 + \tfrac{1}{2}Fe_2O_3 + 2S^0 + H_2O \qquad (1)$$

If the extent to which this reaction occurred was satisfactory and the solid products, iron oxide and elemental sulphur, could readily be removed, this would leave a cupric sulphate solution ready for an electrowinning or hydrogen reduction stage for copper metal. The sulphuric acid would be regenerated and be in balance with the above reaction for recycling to the leach system.

Several attempts to devise processes along the lines of reaction (1) have been made[5] but generally the low yield of this reaction appears to be unacceptable. It has been reported that under optimum conditions of a temperature of 115°C and an oxygen partial pressure of 200–500 psi only 65% of the copper was leached after $2\tfrac{1}{2}$ hours.[25]

The attraction and advantage of a sulphuric acid leach system is quite obvious for leaching sulphide minerals. Further developments have therefore appeared, such as the recent Sherritt–Cominco process which will be discussed below. This will be followed by a description of the main routes in chloride leach systems where the major part of the activity in hydrometallurgical processing seems to be taking place at present. In addition, the use of nitric acid as an oxidising agent will be described and also the later developments in the Arbiter process, where the formation of Chevreul's salt as an intermediate product seem to appear as a common feature with an interesting process making use of acetonitrile as a strong complexing agent for cuprous ions.

3.3   The Sherritt–Cominco process

The Canadian companies, Sherritt Gordon Mines and Cominco Ltd. have in a joint venture developed a process for copper production from sulphide concentrates.[26] The main objective of the project was – as in the case of most projects – to develop a process applicable to a wide range of concentrates. It is interesting to note that one of the reasons for choosing a sulphuric acid system was the potential recovery of precious metals. According to Sherritt–Cominco, a sulphate system rather than a chloride route is to be preferred in this respect. The Au and Ag recovery is claimed to be comparable to smelting practice and the Cu recovery is reported to be 98 %.

At an early stage in the process development it was realised that a serious problem appeared in the form of iron slimes. For this reason a selective removal of the bulk of the iron prior to the oxidation leach of the copper values was prescribed.

This leads to a characteristic and unique feature of the S–C Process: the selective and complete dissolution of iron as one of the primary processing steps. This is considered as equivalent to the converting stage in pyrometallurgical smelting practice for copper concentrates. The analogy may in some respect be justified, in particular on the basis of the subsequent precipitation of the iron as jarosite. Like the slag–forming iron in pyrometallurgical practice, the precipitated jarosites in hydrometallurgy do not seem to have achieved any commercial value. On the contrary the material must usually be disposed of as a deposit which for environmental reasons will tend to increase both the operating and capital costs.

It seems therefore that methods should be sought which would recover and separate the dissolved iron to make a saleable iron product rather than the jarosite precipitate. On the other hand in the case of further oxidation of the elemental sulphur to sulphate, the rejection of iron as jarosite will represent an outlet for sulphate. Thus the precipitation of jarosite may be used to control and keep in balance the sulphate level of the processing circuit. A simplified block flow diagram of the process is given in Fig. 8.

3.3.1 Selective leaching and removal of iron. To achieve the goal of rejecting iron prior to the oxidation leach of the copper, two pretreatment methods representing front-end options were developed in the S–C Process. A *thermal pretreatment* procedure seems to be the most versatile of these, especially for treatment of low-grade concentrates containing pyrite in the chalcopyrite concentrates. This is a two-stage process starting with removal of labile sulphur by partial decomposition of the sulphide minerals:

$$5CuFeS_2 \xrightarrow{\text{Heat}} 5CuFeS_{1.8} + \tfrac{1}{2}S_2 \tag{2}$$

$$7FeS_2 \rightarrow Fe_7S_8 + 3S_2 \tag{3}$$

This is followed by a reduction step with hydrogen:

$$5CuFeS_{1\cdot8} + H_2 \rightarrow Cu_5FeS_4 + 4FeS + H_2S \tag{4}$$

$$Fe_7S_8 + H_2 \rightarrow 7FeS + H_2S \tag{5}$$

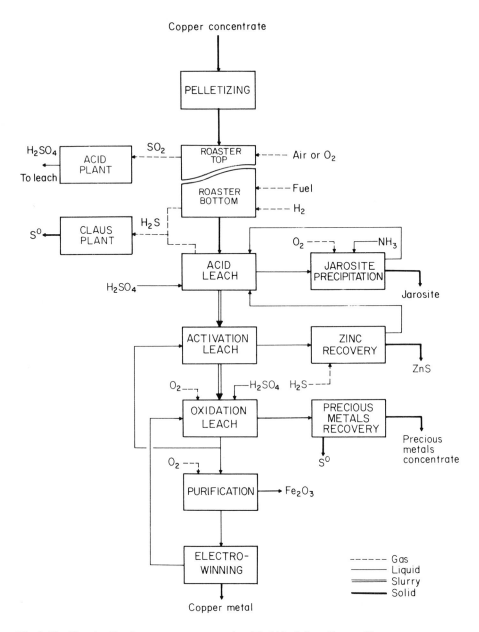

Fig. 8. The Sherritt–Cominco copper process: simplified block flow diagram[26].

The majority of the iron will by this treatment be converted to FeS (troilite) with a minor part of the iron in the copper mineral bornite. The activated calcine is then subjected to an acid leach by which the troilite is readily dissolved:

$$FeS + H_2SO_4 \rightarrow FeSO_4 + H_2S \tag{6}$$

leaving the bornite as a leach residue which is highly reactive in the subsequent oxidation leach of the copper:

$$Cu_5FeS_4 + 6H_2SO_4 + 3O_2 \rightarrow 5CuSO_4 + FeSO_4 + 4S° + 6H_2O \tag{7}$$

The iron dissolved at this stage must be dealt with in the further purification prior to electrowinning of the copper. It has been reported[26] that the solution purification is carried out by a high-temperature oxidation hydrolysis step which removes iron as hematite (Fe_2O_3) scavenging impurities such as Te, As, Bi, Sb, Pb, and Se by coprecipitation with iron oxide. It is also reported that precipitation of iron as jarosite at a lower temperature does not scavenge impurities adequately.

It may be noted that although the amount of iron in solution after leaching the copper is modest, the concentration is far too high (6–10 g/1) for hydroxide precipitation. On the other hand it seems that the introduction of a high-temperature oxidation hydrolysis step for the iron removal must be a complicating and costly operation. As mentioned above, however, the main purpose of this step is purification for effective removal of trace amounts of impurities down to allowable levels which are very strictly enforced. Removal of Se^{6+} is reported to be particularly difficult. Reduction to Se^{4+} has been found necessary prior to the coprecipitation with iron. The reduction step involves heating of the ferrous-containing leach liquors to 200°C under a non-oxidising atmosphere to reduce the Se^{6+} to Se^{4+} by the Fe^{2+}. This is followed by oxidation and hydrolysis of the remaining Fe^{2+} and the coprecipitation of impurities.

In addition to the thermal pre-treatment there is a second front-end option of an *activation leach* reaction:

$$CuFeS_2 + CuSO_4 \xrightarrow{150°C} 2CuS + FeSO_4 \tag{8}$$

This replacement reaction is found to be highly effective for bornite concentrates, less efficient on chalcopyrite, while pyrite does not react. Although the activation leach reaction above is based upon chalcopyrite, the yield of this particular reaction may therefore be rather limited. For further treatment of the bornite product of equation (4), however, an activation leach is bound to work with the main reaction as:

$$Cu_5FeS_4 + CuSO_4 \rightarrow 2Cu_2S + 2CuS + FeSO_4 \tag{9}$$

Taking both activation steps into account, the Sherritt–Cominco Process offers the possibility of a selective leaching and controllable dissolution of iron forming a ferrous sulphate solution. By combining the two front-end options it should be possible to optimise the amount of iron left for the purification step in the subsequent oxidising leach of copper. It is claimed that the activation step should not be too efficient and that a finite amount of iron should be available for purification. However, if the level of iron was kept at 1–2 g/l one might be inclined to think in terms of a purification step similar to that which is practiced for the neutral leach solution in the zinc industry. Normal practice there is to precipitate about 1 g/l Fe as ferric hydroxide which is highly effective for scavenging impurities from the neutral zinc sulphate solution.

3.3.2   Iron rejection. The main part of the iron which was present in the concentrates appears as a ferrous sulphate solution from the acid leach in which the troilite was dissolved. It has been found that rejection of this iron by precipitation of jarosite at a temperature above 60°C is the best choice, adding ammonia so as to form ammonium jarosite. Simultaneously, the ammonia will act partly as a neutralising agent for the acid from the hydrolysis reaction:

$$6FeSO_4 + 2NH_3 + 1 \cdot 5O_2 + 8H_2O \rightarrow 2NH_4[Fe_3(SO_4)_2(OH)_6] + 2H_2SO_4 \quad (10)$$

It is reported[26] that work is in progress to try to separate iron as hematite rather than jarosite. The problems encountered, however, will most likely be connected with the sulphate system itself. It seems generally to be very difficult to produce an iron oxide product with a sulphur content low enough for it to be accepted as a saleable product for iron making. Alternative solutions to this problem, such as solvent extraction of the iron in combination with hydrolysis reactions in the organic phase, which is outlined in the chapter of this volume dealing with zinc extraction, should be worth looking at in connection with upgrading the iron to make a commercial product.

3.3.3   The copper winning circuit. One of the main problems in the copper circuit seems to be the removal of trace elements such as Te and Se. The oxidation leach reaction (7) gives a cupric sulphate solution containing about 100g/l of copper as a basis for depositing copper metal. Sulphur and residual sulphides are separated by flotation, leaving precious metals in the final residue in a concentrated form. As long as a small amount of copper is left in the leaching operations, the precious metals and molybdenum will not be dissolved in the sulphate system. As will be seen from Fig. 8 there are two sulphur-containing gases, coming from the thermal pretreatment and the acid leach and which may be integrated in a Claus plant. Furthermore, a zinc removal step has been included, recovering zinc by precipitation with hydrogen sulphide.

It will be noted that although the S–C Process in many respects represents a pure hydrometallurgical process, the lack of solvent extraction steps, at least as optional alternatives, is striking. The main metals involved – copper, iron, zinc, as well as the many impurities present – should lend themselves to solvent extraction routes. No doubt solvent extraction techniques have been discussed seriously during the process development. The processing flowsheet seems to reflect, however, that companies try, as far as possible, to use the kinds of process with which they have experience, when introducing new methods of treatment which appear to have advantages.[27]

The S–C Process, in its present stage of development, is heavily loaded with a great number of process steps, some of which are very costly, like the high temperature/pressure oxidation hydrolysis operations. No doubt the overall process may work and is technically feasible but the impression of an unfavourable complexity is predominant. The major weakness of the S–C Process seems in several ways to be connected with the iron problem: firstly, due to the iron rejection as the low, or negative, value product jarosite, and secondly, on the grounds of repeating this technique for purification purposes in the oxidation leach circuit. There seems to be a great challenge to try to solve the purification problem by more direct means, for instance by introducing solvent extraction techniques into the processing circuit.

The Sherritt–Cominco staff are to be congratulated for an excellent presentation of their series of papers and for their openmindedness and careful consideration of the problems encountered. It is interesting to note that, in spite of the complexity of the overall process, as well as the difficulties met in solving the practical problems, optimism is clearly expressed by the declaration that 'the S–C Process (is) a viable alternative to the smelting/electrorefining route traditional in the copper industry'.

3.4   The Cymet process

A major attraction of the chloride processes for primary copper production is the possibility of a direct leach operation on the copper minerals at atmospheric pressure. A number of processes starting from this principle have therefore been developed or are presently in progress.[28,29]

A unique approach is demonstrated in the Cymet Process, studied extensively at the pilot plant stage by Cyprus Metallurgical Processes Corporation. The fundamental concept of the Cymet Process is based upon an electrochemical circuit where the sulphide part of the mineral is oxidised to sulphur at an anode with the dissolution of metal ions, and the reduction of these ions to metal at a cathode. Figure 9 presents the flowsheet of the process tested for chalcopyrite at the Cymet plant. Both copper and iron are recovered electrolytically as metal products while the oxidising agent $FeCl_3$ is regenerated simultaneously in the same

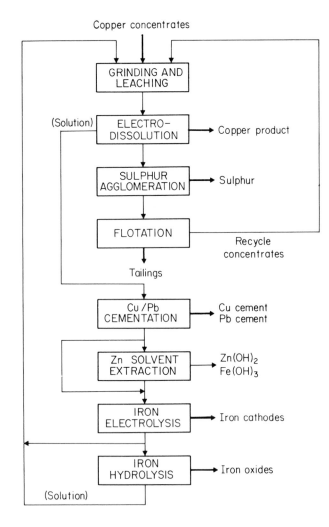

Fig. 9. Diagram of the Cymet process[29].

cells. Although the use of electrical energy is kept to a minimum by the arrangement of the electrochemical circuit, it appears that the economy of the process will be highly dependent on the market price which may be obtained for the iron metal product. It is claimed that a high copper yield of 98 % may be achieved, but it has also to be noted that the copper is precipitated as a powder in the chloride solution and has to go further to an electrorefining step, which will add to the capital and operating costs of the process.

The basic chemistry of the main steps may be represented as follows:
The leach reaction:

$$6FeCl_3 + 2CuFeS_2 \rightarrow 2CuCl + 8FeCl_2 + 4S° \qquad (11)$$

The dissolution takes place in the diaphragm-fitted electrolytic cells, the equations being given as:

$$CuFeS_2 + 3HCl - 3e \rightarrow CuCl + FeCl_2 + 2S° + 3H^+ \qquad (12)$$

$$3CuCl + 3e \rightarrow 3Cu° + 3Cl^- \qquad (13)$$

$$3H^+ + 3Cl^- \rightarrow 3HCl \qquad (14)$$

The spent catholyte solution is further depleted of copper by cementation with metallic iron, followed by a still further purification step by zinc cementation to remove trace elements such as Pb, Sb, Bi and As. The dissolved zinc from this operation is subsequently recovered by solvent extraction with a tertiary amine.

The purified ferrous chloride solution is subjected to electrolysis, making high-purity iron cathodes while ferric chloride is regenerated at the anode:

$$3FeCl_2 + 6e \rightarrow 3Fe° + 6Cl^- \qquad (15)$$

$$6FeCl_2 - 6e + 6Cl^- \rightarrow 6FeCl_3 \qquad (16)$$

The imbalance between cathode and anode efficiencies results in an excess of dissolved iron which is rejected from the processing circuit by hydrolysis and precipitation of iron oxides.

The major problems encountered during the process development are most likely connected with the operation of the electrolytic cells, coupled with a poor quality of the copper powder which will have to undergo further refining. A satisfactory price level for the large amounts of electrolytic iron product would probably be very difficult to achieve for the comparatively large quantities which would be produced.

Apparently, the Cymet process has not been successful in its basic version. An improvement, or rather a new process, has therefore been announced recently[30] as the Cyprus process, while at the same time it was declared that 'Cymet is dead'.

3.5   The Cyprus process

Having learnt the hard way by experience during the Cymet process development, the new chloride process is based on a circuit consisting of more technically conventional steps. The overall flow sheet for processing a pure copper concentrate such as chalcopyrite is shown in Fig. 10.

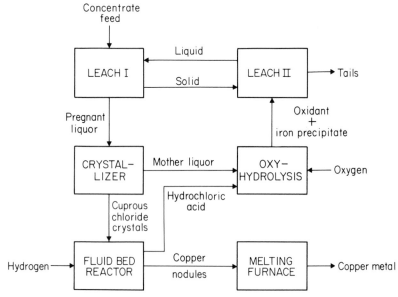

Fig. 10. Flow sheet for the Cyprus copper process.

The concentrate is leached in a two-stage process using a mixed solution of ferric chloride and copper chloride for leaching. In the first stage the copper ions in the initial leach solution are in the cupric state and both the ferric and the cupric ions act as oxidising agents in the leaching process. The basic reactions involved in the first leach stage are:

$$4FeCl_3 + CuFeS_2 \rightarrow 5FeCl_2 + CuCl_2 + 2S° \tag{17}$$

$$3CuCl_2 + CuFeS_2 \rightarrow 4CuCl + FeCl_2 + 2S° \tag{18}$$

$$6CuCl_2 + S° + 4H_2O \rightarrow 6CuCl + H_2SO_4 + 6HCl \tag{19}$$

During the dissolution process, the leach solution will become highly concentrated in cuprous chloride and after being held in a standard thickener for solid/liquid separation the overflow solution is passed to a crystalliser where cuprous chloride is precipitated in a single stage vacuum cooler. After drying, the cuprous chloride is fed along with sand particles into a fluid bed reactor for hydrogen reduction, thereby producing copper metal:

$$CuCl + \tfrac{1}{2}H_2 \rightarrow Cu° + HCl(g) \tag{20}$$

The copper forms around the sand particles and grows in size forming nodules. The copper nodules and sand mixture will be further melted in a conventional furnace. The melting step will include slagging to return the sand to the reactor.

The mother liquor from the crystalliser goes to a hydrolysis step for iron which is controlled by the oxygen supply and where the following reactions take place:

$$4CuCl + O_2 + 4HCl \rightarrow 4CuCl_2 + 2H_2O \tag{21}$$

$$4FeCl_2 + O_2 + 4HCl \rightarrow 4FeCl_3 + 2H_2O \tag{22}$$

$$3FeCl_3 + 2Na_2SO_4 + 6H_2O \rightarrow Na[Fe_3(SO_4)_2(OH)_6] + 3NaCl + 6HCl \tag{23}$$

$$FeCl_3 + 6H_2O \rightarrow Fe(OH)_3(H_2O)_3 + 3HCl \tag{24}$$

The hydrogen chloride supply for reactions (21) and (22) will be in balance with the HCl recovery from the hydrogen reduction step, equation (20), and the acid liberated by the hydrolysis reactions (23) and (24). The product of the iron precipitation is made up of jarosites and hydrated iron oxides, and the oxidants from this oxy-hydrolysis step form a slurry which is passed to the second stage leach. This is essentially a ferric chloride leach and in this stage the remainder of the copper is extracted from the concentrate and converted to cupric chloride. After passing through a thickener the liquor product is returned to the first stage leach and the underflow slurry is sent to the tails filter and then to the tailings pond.

3.5.1   Recovery of by-products and iron rejection. The Cyprus Process looks attractive on the flowsheet and has the simplicity of approach which is a good starting point. However, the discussion of the process so far[30-32] has been limited to the treatment of the pure copper concentrate. It has been stated that by-product recovery is possible and that this has been demonstrated in the laboratory for the recovery of silver, gold, molybdenum and sulphur. When products are to be recovered from the residues, it has to be noted that the tails will contain the appreciable amount of iron precipitated as jarosites and hydrated iron oxides (goethite).

Very little has been mentioned about the iron rejection apart from the hydrolysis reactions (23) and (24). It is believed that the practical operation of the steps for precipitation of jarosites and goethite is likely to represent, at very least, a nuisance in the process circuit. On the other hand it will to a certain extent probably have the advantage of scavenging some of the impurities present.

As an advantage claimed for the process, it has been stated[32] that its operation produces no pollutants in the air or water. This is obviously achieved, however, by including liquid tight areas for impoundment of tailing products. With the large amounts of iron precipitates involved this is likely to add a heavy burden to the capital and operating costs of the process. Nevertheless, it is stated[32] that taking these precautions into account, a feasibility study for the Cyprus process shows that a plant can be built and operated at a lower cost than any known competing process.

3.6   The CLEAR process

Duval Corporation seems to be among the leading companies, if not the leading one, in the application of a chloride system for hydrometallurgical processing of copper concentrates. A 32 500 tons per year plant for commercial operation has been built in Arizona[33] and further extensions are believed to be under consideration.

Apart from the patents[34] not much detailed information has been given about the CLEAR process. The essential feature is, however, that it recovers metallic copper from chalcopyrite and other copper-containing materials by a ferric chloride oxidation leach. The resulting solution of cupric chloride is reduced to cuprous chloride from which copper is recovered by electrolysis in a diaphragm cell.

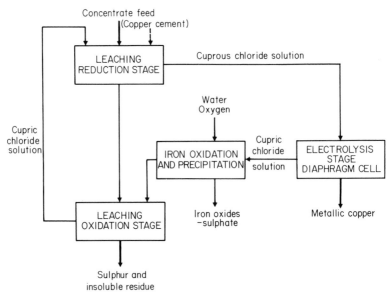

Fig. 11. The CLEAR process[34].

The flowsheet of the process, according to the patent, is shown in Fig. 11. As in the Cyprus process, there are two stages of leaching. What is called the first stage, described as the 'oxidation' stage, is essentially based on the reaction

$$4FeCl_3 + CuFeS_2 \rightarrow 5FeCl_2 + CuCl_2 + 2S° \tag{25}$$

which is identical with the first of the reactions in the Cyprus process. The resulting cupric chloride solution is passed on to the second leaching stage, the 'reduction' stage, where the main purpose is to reduce the cupric ions to the cuprous state, by further reaction with the copper bearing material:

$$3CuCl_2 + CuFeS_2 \rightarrow 4CuCl + FeCl_2 + 2S° \tag{26}$$

Peters[35] has discussed the CLEAR process on the basis of the patent disclosure and suggested that the existing plant probably does not use exclusively chalcopyrite concentrates for the reduction stage, but might utilise more strongly reducing substances, such as cement copper, for part of this step. This feature is also contemplated in the patent where it is stated that the reduction is preferably accomplished in two steps; first, reacting the cupric chloride solution with fresh copper sulphide ores at a controlled temperature, and then with additional reducing agent, such as materials containing metallic copper:

$$Cu° + CuCl_2 \rightarrow 2CuCl \tag{27}$$

To keep the cuprous chloride from precipitating, a suitable amount of sodium chloride is included in the process solution to form the $CuCl_2^-$ anion. The solution is electrolysed in a diaphragm cell where half the copper is deposited at the cathode and the other half is oxidised to cupric chloride. Thus the ideal material balance will be created when half of the copper-bearing material is leached in the 'oxidation' stage (reaction 25) and the other half in the 'reduction' stage (reactions 26 and 27 combined).

It is to be noted that by this ideal balance the electrowinning is a low energy step because the anode oxidises CuCl to $CuCl_2$ while the cathode reduces CuCl to metal at a reversible cell potential of only about 0·4 volts. In practice a somewhat higher potential is necessary to drive the electrodeposition. Peters[35] has pointed out the fact that even with a potential higher by 1 volt, this represents less than one quarter of the energy involved in ordinary electrowinning from copper sulphate solutions.

The copper metal product from the chloride solution electrolysis is most likely to be further refined for making a wire bar quality. It is to be noted that any silver present in the ore is reported to be solubilised as silver chloride from which metallic silver can readily be recovered.

3.6.1   Regeneration and rejection of iron. The regeneration of the ferric leach solution is accomplished by oxidation of the raffinate solution from the electrolysis with either air or oxygen. The presence of cupric chloride catalyses the oxidation of ferrous to ferric chloride.

The necessary bleed of iron from the process circuit is not revealed in detail. From the patent disclosure it seems, however, that once again the approach of precipitation of jarosites and basic iron oxides by hydrolysis is incorporated. Although the advantageous scavenging effects of co-precipitation of metal impurities are secured, the same deposit problems will be met as in the Cyprus process.

3.7   The Minimet Recherche process

A very interesting approach using a cupric chloride solution as the leaching medium, and recovering the copper by solvent extraction – electrowinning has been suggested by Minimet Recherche (Imetal Group).[36] By combining the solvent extraction step and the regeneration of the leach solution, using oxygen as an oxidising agent, only electricity and air are required to operate the process.

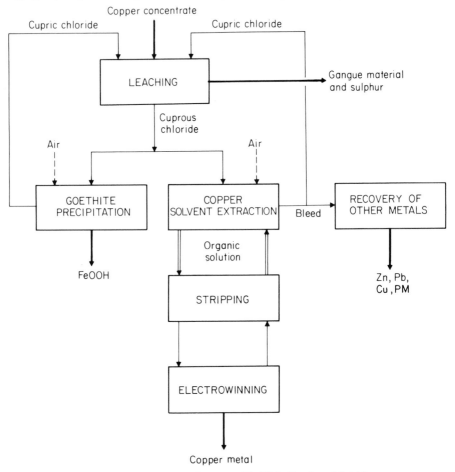

Fig. 12. The Minimet Recherche process for copper[36]. (PM = Precious Metals)

The basic flowsheet is shown in Fig. 12. The leaching step is carried out with a cupric chloride solution containing about 250 g/l of sodium chloride to avoid precipitation of cuprous chloride. The leaching takes place at a pH not greater than 1. For chalcopyrite the temperature is kept close to the boiling point. By leaching in two stages it is reported that more than 98 % of the iron and copper passes into solution, the leaching reaction being:

$$3CuCl_2 + CuFeS_2 \rightarrow 4CuCl + FeCl_2 + 2S° \qquad (28)$$

From this equation it will be seen that only one quarter of the copper from the leach reaction is to be withdrawn from the circuit as a metal product. To achieve this goal the leach solution containing CuCl, FeCl$_2$ and NaCl is split in two halves, with one half directed to the solvent extraction – electrowinning circuit.

The extraction of copper takes place with the copper selective reagent L I X 65N (RH) which will release H$^+$ during the extraction. By a simultaneous introduction of air to the extraction stage, however, the H$^+$ will be consumed by the oxidation of the cuprous ions to the cupric state. Thus the simplified overall reaction for the simultaneous oxidation and extraction will be:

$$2CuCl + 2\overline{RH} + \tfrac{1}{2}O_2 \rightarrow \overline{R_2Cu} + CuCl_2 + H_2O \qquad (29)$$

The loaded solvent may subsequently be stripped by a spent electrolyte solution of sulphuric acid:

$$\overline{R_2Cu} + H_2SO_4 \rightarrow CuSO_4 + 2RH \qquad (30)$$

Both the extraction and the electrolysis steps are closed-loop processes and the copper should be obtained as a high quality product by conventional electrolysis from a sulphate solution.

The aqueous raffinate solution from the solvent extraction, containing regenerated cupric chloride, is returned to the leaching stage, probably via the iron rejection stage for at least partial removal of the appreciable amount of ferrous ions present.

3.7.1   Rejection of iron. From Equation (28) it will be seen that one mole of iron has to be rejected from the processing circuit for each mole of copper produced from chalcopyrite. It is reported that iron will be precipitated as goethite at atmospheric pressure by simple oxidation of the leach solution by air. Both the ferrous and the cuprous ions will thus be oxidised and obviously the coupled effect of the redox systems Cu$^+$/Cu^{2+} and Fe^{2+}/Fe^{3+} is of particular advantage in this case.

The overall reaction for the goethite precipitation will be:

$$2FeCl_2 + 4CuCl + \tfrac{3}{2}O_2 + H_2O \rightarrow 2FeO.OH + 4CuCl_2 \qquad (31)$$

The pH is stabilised at its optimum value of 2·6[37] and the temperature kept above 80°C to ensure that the precipitate has good filterability. It is to be noted that according to Equation (31) the simultaneous cuprous chloride oxidation and goethite precipitation will be in balance with two moles of cuprous ions oxidised for each mole of iron rejected. Taking into account that half of the cuprous to cupric oxidation takes place in the solvent extraction step, it will be realised that the overall Cu/Fe ratio of 4 of the leaching reaction (28) is in balance for the overall process.

The Minimet Recherche Process has been further developed for treatment of complex sulphide ores.[37] It is claimed to be a very flexible process for non-ferrous metals, particularly Zn, Pb and Cu, carried out by selective leaching with the cupric chloride solution. Any pyrite present in the sulphide ore will remain unattacked. Zinc will be recovered from the leaching solution by a similar oxidation–extraction process as is shown above for copper, using an extracting agent selective for zinc.

It seems to be a great advantage of the process that the steps of solvent extraction of copper/zinc and the rejection of iron as goethite can be balanced by the simultaneous oxidation and regeneration of the cupric chloride leach solution, thus avoiding the addition of any further external source of chemical agents within the processing circuit. On the other hand, this may have to be paid for by an appreciable coprecipitation and loss of copper in the goethite precipitation.[37]

The process is reported to have been run and tested successfully on the pilot-plant scale on a number of different types of pyrites. No cost-data seems to have been revealed.

3.8   Processes using nitric acid in sulphate systems

The strong oxidising effect of nitric acid has been known for a long time and processes for leaching copper sulphides by means of nitric acid were examined in the early years of this century.[38] Among the recent developments Björling and his collaborators[39] have suggested a process shown in Fig. 13.

The leaching takes place in two stages with an acid leach stage responsible for about 80% of the copper dissolution. The essential reaction in the acid leaching of chalcopyrite will be:

$$CuFeS_2 + 5H^+ + \tfrac{5}{3}HNO_3 \rightarrow Cu^{2+} + Fe^{3+} + 2S^\circ + \tfrac{10}{3}H_2O + \tfrac{5}{3}NO \qquad (32)$$

Part of the sulphur will be further oxidised to sulphate according to the reaction:

$$S^\circ + 2NO^- \rightarrow SO_4^{2-} + 2NO \qquad (33)$$

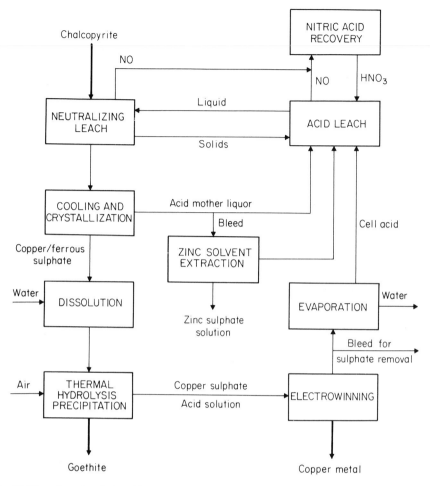

Fig. 13. Flow sheet for nitric acid process[39].

The hydrogen ions are preferentially supplied by sulphuric acid which will be returned from the copper electrolysis cells. It is reported that about 12% of the sulphide sulphur is oxidised to sulphate which will have to be disposed of by liming. At 95°C a retention time of $2\frac{1}{2}$ hours gave almost complete dissolution of the particular sample of chalcopyrite concentrate used. The second leaching stage is actually a combination of neutralisation of the excess nitric acid and reduction of the ferric ions to ferrous. Sulphuric acid (cell acid) will be added to keep the acidity high enough for reaction (32) to proceed.

The nitric oxide gas from the leaching is captured and reconverted to nitric acid by oxidation with air and absorption in an aqueous solution according to the reaction:

$$2NO + 3/2O_2 + H_2O \rightarrow 2HNO_3 \tag{34}$$

After the second leaching stage the separation of iron and copper from the leach solution is done by co-crystallisation of cupric and ferrous sulphates by cooling to 5°C. Any nitrates left will remain in the mother liquor which is returned to the acid leach.

The sulphate crystal mixture is dissolved in water and iron is oxidised and rejected as goethite in an autoclave at an air pressure of 10 atmospheres and a temperature of 160°C. Some co-precipitation of copper is unavoidable but after washing with dilute nitric acid a final content of 0·55% Cu in the calcined goethite is reported. The iron-free solution is taken to copper electrowinning and the cell acid recirculated to the leaching circuit after partial evaporation. Total recovery of copper is reported to be about 96%.

The flowsheet shows an outlet for zinc which is said to be removed from the mother liquor by solvent extraction. No details are revealed about the extracting agent used, although it is reported that zinc will be extracted together with copper and iron and the zinc finally recovered by electrowinning. There is no doubt that this will be technically possible, but it is believed that the capital and operating costs for the zinc recovery will be rather high.

The crystallisation and subsequent dissolution of the mixture of copper and ferrous sulphates followed by the autoclave oxidation and precipitation of goethite and also the partial evaporation of water from the cell acid, all seem to be stages which are likely to create difficulties for an economically viable process. These steps have been avoided in the Nitric-Sulphuric Leach (NSL) process developed by E.I. duPont de Nemours Co and further evaluated by Kennecott Copper Corp.[40,41] The flowsheet of the NSL process is shown in Fig. 14. It is claimed that it is an improvement over previous nitric acid processes, including the systems described by Björling et al.[39] (Fig. 13) and also Prater et al.[42], in providing the advantages of:

● copper electrowinning from leach liquor
● elimination of leached tails refloat
● recovery of molybdenum and precious metals.

The leaching was originally carried out in a three-stage counter-current arrangement with solid-liquid separation between each stage. This was later simplified to two stages.[41] It is reported that acceptable reaction rates were achieved in the range 90 to 105°C which avoided the use of pressure reactors and allowed the use of brick lined vessels with stainless steel trim. Reactions (32) to (34) will all be used in the leaching circuit. In the improved design of the leach and nitric acid recovery steps, NO is regenerated as NO_2 by reacting with oxygen and the NO_2 product is sparged directly into the leach reactor.

In the leach step, essentially all of the metal sulphides are dissolved. If molybdenum is present, it will be removed from the leach liquor by solvent extraction. The pregnant liquor is subsequently fed into a double autoclave system which

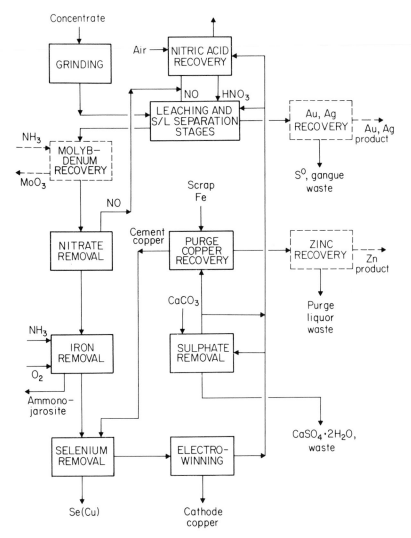

Fig. 14. Nitric–sulphuric leach flow sheet[40].

serves two purposes, firstly – residual nitrate is reduced to NO while Fe^{2+} is simultaneously oxidised to Fe^{3+} and secondly – iron is precipitated and rejected partially as hydrogen jarosite $H[Fe_3(SO_4)_2(OH)_6]$ in the first autoclave and as ammonium jarosite, $NH_4[Fe_3(SO_4)_2(OH)_6]$ in the second autoclave, where a small amount of oxygen may be added to oxidise any residual Fe^{2+}.

It is interesting to note that at this stage the Nitric-Sulphuric Leach process is faced with the same problem as the Sherritt–Cominco process with regards to selenium which will be present as Se^{6+}. In the present process Se^{6+} is reduced to

Se^{4+} and precipitated by contacting with recycled cement copper, which in some respects seems more simple than the autoclave treatment at 200°C in the S–C process. On the other hand it will be remembered that the autoclave treatment in the S–C process was a combined procedure in which the Se^{6+} was reduced by Fe^{2+} and the excess Fe^{2+} was subsequently oxidised and precipitated either as jarosite or preferably as hematite.

The purified pregnant liquor is fed to electrowinning cells where cathode grade copper is recovered. The sulphate level is controlled by precipitation of gypsum from the depleted electrolyte. The major part of the sulphate rejection, however, will take place by the precipitation of jarosites. The extent of oxidation of sulphur to sulphate by reaction (33) depends on the mineralogy of the sulphide concentrate. It is reported that under normal leach conditions the conversion of sulphide to $S°$ will be about 65% for chalcopyrite ($CuFeS_2$). Assuming that the remaining 35% is oxidised further to sulphate, these figures will be in close balance with the rejection as jarosite – $NH_4[Fe_3(SO_4)_2(OH)_6]$ – where the sulphur/iron ratio is one-third of the ratio in the chalcopyrite mineral.

Although the precipitation of jarosite in the NSL process seems to be a sensible solution to both the iron and the sulphur problem in this particular case, it still is believed to be a disadvantage of the process. This is primarily connected with the general deposition problem created by the jarosite itself. A severe co-precipitation of copper is most likely to occur, representing a copper loss, apart from the environmental precautions to be taken. Furthermore, the addition of ammonia and the sulphide oxidation to sulphate will be reflected in the operating costs.

Nevertheless, the economic analysis and evaluation, including the improved design of the leach and nitric acid recovery steps and the use of air agitation electrowinning in the tank house,[41] indicates that the NSL process is a competitive process among the hydrometallurgical processes offered on the market. It is also claimed to compare favourably with flash smelting for plants smaller than about 50 000 tons per year of copper. The evaluations so far are, however, based upon laboratory results and it is stated[40] that commercialisation will require a pilot plant programme in which chemistry, control strategy, equipment scaleup and materials of construction will have to be demonstrated for several months of operation.

3.9   The Arbiter process

On the basis of the earlier Sherritt Gordon processes for high pressure leaching using ammonia as a leaching agent for sulphide concentrates, the Anaconda Company has developed the Arbiter process where essentially the same reactions are taking place:

$$2CuFeS_2 + 8\tfrac{1}{2}O_2 + 12NH_3 + 2H_2O$$
$$\rightarrow 2Cu(NH_3)_4SO_4 + 2(NH_4)_2SO_4 + Fe_2O_3 \qquad (35)$$

In practice the process does not work efficiently with chalcopyrite itself. The main feature of the Arbiter process is that the leaching takes place at atmospheric pressure with pure oxygen in specially designed leaching equipment. In this, vigorous agitation is applied to provide adequate mass transport of oxygen to the mineral/solution interface for oxidation of the mineral[43].

The copper may be recovered from the leach liquor by solvent extraction and there are two options available for the disposal of the ammonium sulphate: either by crystallisation as a by-product or by adding lime for recovery of ammonia and discarding the sulphate as gypsum. The original flowsheet for the Arbiter process is shown on Fig. 15.

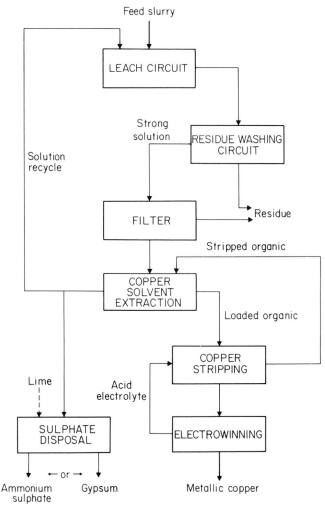

Fig. 15. The Arbiter process[44].

In order to replace the high cost operation of electrowinning, investigations have been in progress looking for other methods for metal production from the copper tetra-ammine sulphate solutions obtained by the ammonia leaching.[44,45]

These investigations have resulted in an interesting alternative to the original Arbiter process. By treating the ammonia leach solution with sulphur dioxide several intermediate products can be obtained by precipitation, of which the main ones are:

- Chevreul's salt: $Cu_2SO_3.CuSO_3.2H_2O$
- Monocupric hepta cuprous
  triammonium sulphite: $Cu_8(NH_4)_3(SO_3)_6.12H_2O$
- Cuprous ammonium sulphite: $Cu_2SO_3(NH_4)_2SO_3$

Under specific conditions the preferred intermediate product, $CuNH_4SO_3$, will be obtained at operating temperatures during precipitation below 60°C. By thermal decomposition of the cuprous ammonium sulphite in an autoclave at 140°–170°C, nearly 100% copper yields as high-purity copper powder have been obtained.

The simplified reactions involved in the reduction and decomposition will be:

$$2Cu(NH_3)_4SO_4 + 3SO_2 + 4H_2O \rightarrow 2CuNH_4SO_3 + 3(NH_4)_2SO_4 \tag{36}$$

$$2CuNH_4SO_3 \rightarrow 2Cu° + (NH_4)_2SO_4 + SO_2 \tag{37}$$

The copper powder product is claimed to be of high purity due to the three stages of purification, i.e. the selective ammonia leaching followed by the two selective precipitations with intermediate washings. It is further reported that the general feasibility of the process has been demonstrated by successful operation of a pilot plant in continuous flow, during which no significant problems were experienced.[45] A schematic flowsheet of the precipitation process is shown in Fig. 16.

3.10   Acetonitrile as complexing agent for copper

Acetonitrile (CH_3CN) is soluble in water. It also forms a strong complex with cuprous ions. In an aqueous solution of copper sulphate the copper ions will normally exist only in the cupric state. However, in the presence of acetonitrile, cuprous sulphate solutions will be stabilised when the aqueous acidic solution (pH < 5) contains 4 moles of acetonitrile per mole of cuprous ion. Fairly concentrated solutions, containing as much as about 90 g/l of Cu^+ may be prepared.

When this solution is heated the volatile acetonitrile will be distilled off and copper metal precipitated according to the disproportionation reaction:

$$Cu_2SO_4 \rightarrow Cu° + CuSO_4 \tag{38}$$

These characteristic features of the acetonitrile – water system have been made use

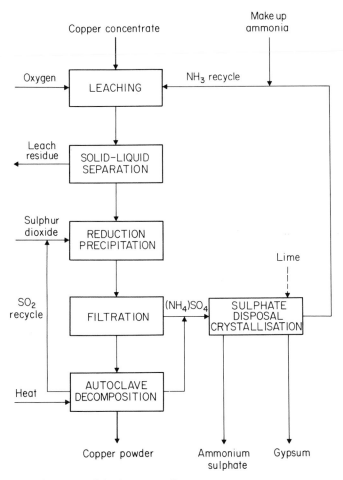

Fig. 16. The Anaconda SO$_2$ precipitation process[45].

of in various process routes developed at the Mineral Chemistry Research Unit at Murdoch University in Australia.[46,47]

Several methods of approach are possible. Cupric ions may, for instance, be used as the oxidising agent for sulphide ores, dissolving copper as cuprous ions in the presence of acetonitrile. A combination with the Arbiter process (the SO$_2$ precipitation alternative) appears interesting, by treating the copper sulphites (Chevreul's salt) with excess of cupric sulphate in acetonitrile-water to give a solution of cuprous sulphate by the reaction:

$$3CuSO_4 + Cu_2SO_3.CuSO_3.2H_2O \rightarrow 3Cu_2SO_4 + 2H_2SO_4 \qquad (39)$$

It is reported[46] that this reaction is limited to a concentration of about 50 g/l Cu$^+$. The acidic cupric sulphate solution and the acetonitrile-water azeotrope will be recycled, but the sulphuric acid produced by the above reaction obviously needs to be neutralised.

3.11   Meeting the iron problem by techniques of solvent extraction

In the hydrometallurgical processes for treating copper concentrates discussed in this review some emphasis has been placed on the iron problem and the short-comings of the methods of solving it by precipitation of the various kinds of iron oxides (goethite, hematite) or basic iron salts like the jarosites, from the acid leach liquors. Coprecipitation of impurities like As, Sb, Te, Se etc. will undoubtedly have an advantageous overall effect, but the main problem is the copper itself which will contaminate the iron product and make it unsuitable for further use such as for instance in the steel-making industry. Since goethites and jarosites are not entirely stable against weathering, the copper content may also create difficulties in disposal of them as waste products.

To this author's knowledge hardly anyone has seriously discussed the possibility of solving the iron problem in copper hydrometallurgy by the techniques of solvent extraction. The reason for this is obvious: because of the recognised high-cost operation of solvent extraction it is quite rightly considered prohibitive if normal techniques of solvent extraction are used.

In some special hydrometallurgical processes it is in use, however, for example in the Falconbridge matte leach process[48] where iron present as an impurity is extracted from strong nickel chloride solution with tributyl phosphate (TBP). This leach solution is, however, completely different from the normal copper leach solutions originating from chalcopyrite concentrates where the quantity of dissolved iron normally will be higher than the amount of dissolved copper.

Spitzer[49] has considered the use of carboxylic acids as extractants in copper hydrometallurgy. However, pointing to the fact that one molecule of alkali is consumed per valency of the iron removed, he came to the conclusion that: 'In the case of trivalent iron – which is invariably present in copper leach liquors – this means a considerable caustic consumption, which has been shown by calculation to make carboxylic acid extraction often economically unattractive for this type of separation.'

The prohibitive feature of the need to add alkali to the processing circuit can be avoided, however, by the technique outlined in recent patents by the present author.[50] In principle the alkaline property of the metal-bearing raw material itself is used to advantage. This has particularly been shown to be suitable for zinc hydrometallurgy[50,51] and the method is described in the chapter of this volume dealing with the Electrolytic Production of Zinc. It may also, with appropriate modifications, be suitable for the treatment of copper solutions.

4 References

1 Biswas A.K. and Davenport W.G. *Extractive Metallurgy of Copper*. Oxford, etc.: Pergamon, 1976, 438 p.

2 Anon. Facing the change in copper technology. *Eng. Min. J.*, Apr. 1973, **174**, A-HHH or *Chem. Eng.*, Apr. 16, 1973, **80**, A-HHH.

3 Yannopoulos J.C. and Agarwal J.C. eds. *Extractive Metallurgy of Copper, vol. I–II.* Int. Symp. Copper Extr. Ref., Las Vegas 1976. New York: AIME, 1976, 1055 p.

4 Jones M.J. ed. *Copper Metallurgy: Practice and Theory.* London: Institution of Mining and Metallurgy, 1975, 83 p.

5 Paynter J.C. A review of copper hydrometallurgy. *J.S. Afr. Inst. Min. Metall.*, 1973, **74**, 158–70.

6 Yazawa A. Trends in modern copper smelting processes. *Erzmetall*, 1977, **30**, 511–7.

7 Subramanian K.N. and Jennings P.H. Review of the hydrometallurgy of chalcopyrite concentrates. *Can. metall. Q.*, 1972, **11**, 387–400.

8 Steintveit G., Lindstad T. and Tuset J.K. Smelting of lead-silver residue and jarosite precipitate. In *Advances in Extractive metallurgy 1977.* London: Institution of Mining and Metallurgy, 1977, 7–12.

9 Agarwal J.C. and Yannopoulos J.C. Copper introduction and overview. In *Extractive Metallurgy of Copper.* New York: AIME, 1976, vol. I, xv–xxiii.

10 Härkki S., Aaltonen O. and Tuominen T. High grade matte production with oxygen enrichment by the Outokumpu flash smelting method. In *Extractive Metallurgy of Copper*, Yannopoulos J.C. and Agarwal J.C. eds. New York: AIME, 1976, vol. I, 488–507.

11 Bell M.C., Blanco J.A., Davies H. and Sridhar R. Oxygen flash smelting in a convertor, *J. Met.*, Oct. 1978, **30**, 9–14.

12 Schuhmann R.Jr. and Queneau P.E. Thermodynamics of the Q–S oxygen process for copper-making. In *Extractive Metallurgy of Copper*, Yannopoulos J.C. and Agarwal J.C. eds. New York: AIME, 1976, vol. I, 76–89.

13 Mills L.A., Hallett G.D. and Newman C.J. Design and operation of the Noranda process continuous smelter. In *Extractive Metallurgy of Copper*, Yannopoulos J.C. and Agarwal J.C. eds New York: AIME, 1976, vol. I, 458–87.

14 Ammann P.R., Kim J.J., Crimes P.B. and Brown F.C. The Kennecott slag cleaning process. In *Extractive Metallurgy of Copper*, Yannopoulos J.C. and Agarwal J.C. eds. New York: AIME, 1976, vol. I, 331–50.

15 Nagano T. and Suzuki T. Commercial operation of Mitsubishi continuous copper smelting and converting process. In *Extractive Metallurgy of Copper*, Yannopoulos J.C. and Agarwal J.C. eds. New York: AIME, 1976, vol. I, 439–57.

16 Amsden M.P., Sweetin R.M. and Treilhard D.G. Selection and design of Texasgulf Canada's copper smelter and refinery. *J. Met.*, July 1978, **30**, 16–26.

17 Mackey P.J., McKerrow G.C. and Tarassoff P. Minor elements in the Noranda process. Paper presented at 104th annual meeting of AIME, New York, Feb. 1975.

18 Burkin A.R. *Int. Metall. Rev.*, Sept. 1976, No. 207, 118–23.

19 Flett D.S. Solvent extraction in copper hydrometallurgy: a review. *Inst. Min. Metall., Trans., Sect. C*, 1974, **83**, C30–8.

20 Flett D.S. Solvent extraction in hydrometallurgy. *Chem. Ind. (London)*, 1977, 706–12.

21 Van der Zeeuw A.J. Liquid/liquid extraction of copper and nickel with selective reagent. *Erzmetall*, 1977, **30**, 139–45.

22 Tumilty J.A., Dalton R.F. and Massam J.P. The Acorga P-5000 series: a novel range of solvent-extraction reagents for copper. In *Advances in extractive metallurgy 1977.* London: Institution of Mining and Metallurgy, 1977, 123–31.

23 Hopkins W.R. and Lynch A.J. Anamax oxide plant: a new US dimension in solvent extraction. *Eng. Min. J.*, Feb. 1977, **178**, 56–64.

24 Fisher J.F.C. and Notebaart C.W. Metallurgical treatment of Chingola cupriferous mica ores. *Inst. Min. Metall., Trans., Sect. C*, 1976, **85**, C15–22.

25 Vizsolyi A., Veltman H., Warren I.H. and Mackiw V.N. Copper and elemental sulphur from chalcopyrite by pressure leaching, *J. Met.*, Nov. 1967, **19**, 52–9.
26 (a) Swinkels G.M. and Berezowsky R.M.G.S. The Sherritt–Cominco copper process · Part I: the process. *CIM Bull.*, Feb. 1978, **71**, 105–21.
   (b) Kawulka P., Kirby C.R. and Bolton G.L. The Sherritt–Cominco copper process – Part II: pilot-plant operation. *CIM Bull.*, Feb. 1978, **71**, 122–30.
   (c) Maschmeyer D.E.G., Milner E.F.G. and Parekh B.M. The Sherritt–Cominco copper process – Part III: commercial implications. *CIM Bull.*, Feb. 1978, **71**, 131–9.
27 Burkin A.R. The winning of non-ferrous metals, 1974. *Proc. R. Soc. London, Ser. A*, 1974, **338**, 419–37.
28 Dutrizac J.E. and MacDonald J.C. Ferric iron as a leaching medium. *Miner. Sci. Eng.*, Apr. 1974, **6**(2), 59–100.
29 Kruesi P.R., Allen E.S. and Lake J.L. Cymet process – hydrometallurgical conversion of base-metal sulphides to pure metals. *CIM Bull.*, June 1973, **66**, 81–7.
30 Anon. New copper process from Cyprus is billed as 'technological breakthrough'. *Eng. Min. J.*, Oct. 1977, **178**, 33.
31 Anon. Cyprus reveals details of new copper process. *Eng. Min. J.*, Nov. 1977, **178**, 30–33.
32 McNamara J.H., Ahrens W.A. and Franek J.B. A hydrometallurgical process for the extraction of copper. Paper presented at 107th annual meeting of AIME, Denver, Feb. 1978.
33 Anon. Duval building 32,500-tpy CLEAR process plant. *Eng. Min. J.*, June 1976, **177**, 245.
34 Atwood G.E. and Curtis C.H. for Duval Corporation. U.S. Patent 3 785 944, Jan. 1974.
35 Peters E. Applications of chloride hydrometallurgy to treatment of sulphide minerals. In *Chloride hydrometallurgy*. Brussels: Benelux Metallurgie, 1977 1–36.
36 Demarthe J.M., Gandon L. and Georgeaux A. A new hydrometallurgical process for copper. In *Extractive metallurgy of copper*, Yannopoulos J.C. and Agarwal J.C. eds. New York: AIME, 1976, vol. II, 825–48.
37 Demarthe J.M. and Georgeaux A. Hydrometallurgical treatment of complex sulphides. In *Complex metallurgy '78* Jones M.J. ed. London: Institution of Mining and Metallurgy, 1978, 113–20.
38 Roman R.J. and Benner B.R. The dissolution of copper concentrates. *Miner. Sci. Eng.*, Jan. 1973, **5** (1), 3–24.
39 Björling G., Faldt I., Lindgren E. and Toromanov I. A nitric acid route in combination with solvent extraction for hydrometallurgical treatment of chalcopyrite. In *Extractive metallurgy of copper*, Yannopoulos J.C. and Agarwal J.C. eds. New York: AIME, 1976, vol. II, 725–37.
40 Brennecke H.M., Bergmann O., Ellefson R.R., Davies D.S., Lueders R.E. and Spitz R.A. The nitric sulfuric leach process for recovery of copper from concentrate. Paper presented at 107th annual meeting of AIME, Denver, Feb. 1978.
41 Davies D.S., Lueders R.E., Spitz R.A. and Frankiewicz T.C. Nitric-sulfuric leach process improvements. Paper presented at 107th annual meeting of AIME, Denver, Feb. 1978.
42 Prater J.D., Queneau P.B. and Hudson T.J. Nitric acid route to processing copper concentrates. *Trans. Soc. Min. Eng. AIME*, 1973, **254**, 117–22.
43 Kuhn M.C., Arbiter N. and Kling H. Anaconda's Arbiter process for copper. *CIM Bull.*, Feb. 1974, **67**, 62–73.
44 Arbiter N., Milligan D. and McClincy R. Metal production from copper ammine solution with sulfur dioxide. In *Hydrometallurgy*, Davies G.A. and Scuffham J.B. eds. London: Institution of Chemical Engineers, 1975, 1.1.–1.9. (*Symposium series* no. 42).
45 Arbiter N. and Milligan D.A. Reduction of copper ammine solutions to metal with sulfur dioxide. In *Extractive Metallurgy of Copper*, Yannopoulos J.C. and Agarwal J.C. eds. New York: AIME, 1976, vol. II, 974–93.
46 Muir D.M. and Parker A.J. In *Advances in Extractive Metallurgy 1977*. London: IMM, 1977, 191–5.
47 Parker A.J. and Muir D.M. Copper from copper concentrates via solutions of cuprous sulfate in acetonitrile-water solutions. In *Extractive Metallurgy of Copper*, Yannopoulos J.C. and Agarwal J.C. eds. New York: AIME, 1976, vol. II, 963–73.

48 Wigstöl E. and Fröyland K. Solvent extraction in nickel metallurgy – the Falconbridge matte leach process. In *Solvent Extr. Metall. Processes*, Proc. Int. Symp., Antwerp, May 1972, 62–72.

49 Spitzer E.L.T.M. The use of organic chemicals for the selective liquid/liquid extraction of metals. In *Solvent Extr. Metall. Processes*, Proc. Int. Symp., Antwerp, May 1972, 14–8.

50 Thorsen G. U.S. Patent 4 008 134, Feb. 1977 and Brit. Pat. 1 474 944, May 1977.

51 Thorsen G. and Grislingås A. Solvent extraction of iron in zinc hydrometallurgy. Paper presented at 109th annual meeting of AIME, Las Vegas, Feb. 1980.

The extractive metallurgy of deep-sea manganese nodules

A. J. Monhemius

1   Introduction

The enigmatic potato-sized lumps of mixed manganese and iron oxides, popularly known as manganese nodules, were first discovered, lying on the bed of the Pacific Ocean, during the voyage of *HMS Challenger* just over one hundred years ago. To this day, the origin of manganese nodules remains a subject of controversy, but it is apparent that the phenomenon is widespread. Nodule deposits have been found in the North and South Atlantic and the Indian Ocean, as well as in the South Pacific.

Manganese nodules remained a scientific curiosity until the early 1960s, but since that time, serious commercial interest has been shown in them as a potential source of certain non-ferrous metals, which occur as minor components of the nodules. One reason for this interest is the vast sizes of the nodule deposits. Those in the Pacific Ocean, for example, have been estimated to be of the order of 10^{11}

tons[4]. Metals which, in the future, may be recovered from manganese nodules include nickel, copper, cobalt, molybdenum and, possibly, manganese itself.

Several industrial consortia have been engaged in research and development into the various aspects of metals recovery from deep-sea manganese nodules. The most technologically daunting task is that of mining or 'harvesting' the nodules, which lie on the ocean floor at depths of 2000 to 3000 fathoms. Numerous mining systems have been proposed and some prototype dredges have been and are being tested, but full-scale mining has not yet begun.

Apart from the technological problems of deep-ocean mining, which have yet to be overcome, two other major problems have to be resolved before commercial exploitation can take place. The first is political and the second environmental.

The political difficulties arise because there is no international law governing ownership of nodule deposits. In view of the inconclusive nature of the recent Law of the Sea Conference, it does not seem likely that this question will be settled by international agreement in the near future. Until the legal situation is clarified, and in the short term, this may be done by unilateral declarations of ownership by individual nations, the very large financial investments needed to make deep-ocean mining a reality are unlikely to be made.

The environmental problem stems mainly from the fact that the dredging operations may result in large quantities of very fine silt from the ocean floor, together with cold, nutrient-rich, deep-ocean water being discharged into the surface waters. Serious concern has been voiced about the effects of these disturbances on the ecology of the oceans. The potential effects are, of course, extremely difficult to predict, and will depend to a large extent on the types of mining systems eventually used for full-scale production. This concern is unlikely to prevent the first generation of ocean mining ships from going into operation. However, if the ecological damage caused by these is seen to be serious, then the long-term future of deep-ocean mining may be seriously jeopardised.

Compared with the difficulties involved in mining manganese nodules, the problem of processing them to extract the valuable metals appears relatively straightforward. Nevertheless, deep-sea manganese nodules are quite unlike any terrestrial ores, both with respect to their physical characteristics and to their mineralogical and chemical compositions and therefore new processes are required. The unique nature of nodules has an overriding influence on the methods that can be used economically to recover metals from them and a brief summary of their chemical and physical properties is necessary before the extractive metallurgy of nodules processing is discussed.

The chemical compositions of nodules from different locations vary widely. Pacific Ocean nodules generally contain the highest concentrations of valuable metals and are the most studied. The average composition of Pacific Ocean nodules, taken from over fifty locations, is shown in Table 1.

Mineralogically, the nodules consist mainly of an intimate mixture of manganese and iron oxides. The grain size is extremely small, ranging from less than one micrometre to about five micrometres. Manganese may be present in up to three distinct mineral phases, whereas iron is present generally only as goethite. The main oxide minerals are listed in Table 2.

Table 1. Average composition of Pacific Ocean Nodules[1,2]
(Concentrations in wt %)

Valuable Metals: Mn, 24·2; Ni, 0·99; Cu, 0·53; Co, 0·35; V, 0·054; Mo, 0·052; Zn, 0·047

Impurities: Major: Fe, 14·0; Si, 9·4; Al, 2·9; Na, 2·6; Ca, 1·9; Mg, 1·7

Minor: K, Ti, Ba, P, S, 1·0–0·1; Pb, Sr, Zr, B, Y, La, C, 0·1–0·01; Yb, Cr, Ga, Sc, Ag, As, Sb, < 0.01

Table 2. Mineralogical Composition of Manganese Nodules[3]

Mn *Minerals:*   Todorokite (Ca, Na, Mn(II), K) (Mn(IV), Mn(II), Mg)$_6$O$_{12}$.3H$_2$O
Birnessite (Na$_7$Ca$_3$)Mn$_{70}$O$_{140}$.28H$_2$O
Delta manganese dioxide δ-MnO$_2$

Fe *Minerals:*   Goethite α-FeOOH

The valuable metals in the nodules appear to be present as an integral part of the manganese and iron oxides, due either to lattice substitution, ion exchange, or adsorption. The porosity and internal surface area of the nodules are very high. Porosities are usually greater than 50%, with pore size diameters in the range 0·1 to 0·01 micrometres, and surface areas are of the order of 200–300 m²/gm. Due to the high porosity, raw nodules generally contain 30–40% sea water, together with its contained salts.

The combination of physical and chemical characteristics of nodules make impossible the application of methods of physical beneficiation to produce concentrates of the valuable metals, and chemical processing has to be used.

In order to obtain high recoveries of the valuable metals, it is necessary to release them by disruption of the crystal lattices of the manganese oxides. This can be achieved by reduction of manganese from the tetravalent to the divalent state. Reduction can be accomplished pyrometallurgically by smelting with gaseous, liquid, or solid reductants. The objective of reduction smelting is to produce a slag containing all the manganese and the majority of the impurities, apart from iron, while the valuable metals are concentrated and recovered, together with most of the iron, in a metallic alloy phase. Alternatively, hydrometallurgical processing can be used, with reduction being carried out either prior to or during leaching. Prior reduction may be accomplished by roasting the nodules in a reducing atmosphere, such as H$_2$, CO or SO$_2$, or by roasting nodules mixed with solid reductants, such as coal.

The fact that raw nodules contain more than 30% moisture, which has to be removed during high temperature processing, is a major disadvantage for both smelting and reduction roasting routes. Much attention has therefore been directed towards reduction leaching processes in which the only nodule pre-treatment required is comminution. Conditions under which manganese is reduced to the divalent state during leaching can be created by using reagents such as SO_2, CO, HCl or $FeSO_4$.

Most of the research and development work directed towards complete processes for the recovery of metals from manganese nodules has been carried out by industrial concerns and information concerning the majority of this work is available only from the patent literature. The greater part of the original work that has been published in the open scientific and technical literature is concerned mainly with the fundamental leaching behaviour of nodules. This work has been reviewed by several authors[3-5] and will not be further considered here. Similarly, economic aspects of metals production from nodules, which has been the subject of a number of papers[5-7], will not be dealt with. In contrast, the patent literature on nodules processing has previously been reviewed only sketchily. The most comprehensive coverage of the patents relating to the solvent extraction treatment of liquors arising from nodule leaching is in the reviews by Flett.[8,9] The present review concentrates on the information available in the patent literature, and, in particular, on patents assigned to two companies, Kennecott Copper Corporation and Deepsea Ventures Incorporated. These two companies dominate the patent literature on manganese nodule processing, with both companies each holding well over twenty patents in this field.

Taken together, the work of Kennecott and Deepsea Ventures covers many of the options available for nodule processing. There are many interesting similarities and some fundamental differences in their approaches to the problem. Both use basically hydrometallurgical routes to recover the valuable metals, with solvent extraction as the key unit process for metals separation, and electrowinning as the main metal recovery option. The most important difference in approach is that Deepsea Ventures aim to recover not only the valuable metals but also manganese from the nodules and all their processes are designed to dissolve manganese in the leach liquors. Kennecott, on the other hand, base their processes on the recovery of nickel, copper, cobalt and molybdenum only and manganese is rejected with the solid leach residues, although the option exists to recover manganese from the residues if market conditions make it favourable to do so. Deepsea Ventures have investigated both chloride-based and sulphate-based routes, with the emphasis on the former, whereas Kennecott have concentrated almost exclusively on ammonia-based routes.

The work of two other companies is also included in this review. These are the International Nickel Company (INCO) and Metallurgie Hoboken-Overpelt

(MHO). Both these companies have published much less extensively than either Kennecott or Deepsea Ventures, but nevertheless, their processes illustrate alternative approaches to the problems of nodule processing. MHO have developed a direct leach chloride-based process, which overcomes many of the problems which beset the Deepsea Ventures chloride-based routes. INCO, on the other hand, is the only company of those considered, which has opted for a smelting route to upgrade the valuable metals contained in the nodules.

The review concludes with an attempt to compare and contrast the technical merits of these various approaches to manganese nodule processing.

2   Deepsea Ventures Processes

The approach taken by Deepsea Ventures (DSV) to the processing of manganese nodules is unique in one important aspect – in addition to nickel, copper and cobalt, the processes are designed also to recover manganese from the nodules. To do this, chloride-based routes appear to be favoured, although DSV have taken patents on several sulphate-based routes.

Most of the processes produce leach liquors containing Mn, Ni, Cu and Co in chloride or sulphate solutions. In the case of chloride solutions, iron may also be present in the leach liquor. These metals are separated by a solvent extraction process which appears to be fairly well developed and standardised. The same process appears, with minor variations, in at least six patents assigned to Deepsea Ventures and it is applicable to both chloride and sulphate liquors. With chloride liquors, however, iron can be readily extracted with amine extractants and therefore can be tolerated in the pregnant leach liquors, whereas for sulphate liquors, iron extraction is more difficult and the DSV sulphate processes are designed to reject iron into the solid leach residues.

Most of the many other DSV patents are concerned either with the front-end operations, i.e. nodule pre-treatment and leaching operations, or with methods of recovering manganese, as metal or oxide, from the manganese chloride liquors produced by the solvent extraction process.

Thus in reviewing the information available from patents assigned to Deepsea Ventures, it is convenient to start with the solvent extraction process, which lies at the heart of many of the flowsheet options. Following this, the patents dealing with front-end and rear-end operations will be discussed.

2.1   Solvent extraction

For chloride solutions, the Deepsea Ventures solvent extraction process consists of five basic steps.[15,18,20,21]

(i)   Selective extraction of iron with a secondary amine, followed by stripping with dilute acid to produce a ferric chloride solution.

(ii)   Selective extraction of copper with LIX 64N, followed by stripping with return electrolyte to produce a copper sulphate advance electrolyte.

(iii)   Co-extraction of nickel and cobalt with Kelex 100.

(iv)   Selective stripping of nickel from the loaded Kelex with return electrolyte to produce a nickel chloride or sulphate solution.

(v)   Selective stripping of cobalt with strong HCl to produce an acidic cobalt chloride solution. This is further treated by re-extraction of cobalt into tri-iso-octylamine, followed by stripping with return electrolyte, to produce a cobalt chloride solution suitable for electrowinning.

In the case of sulphate leach liquors, iron is not present in the feed to solvent extraction, having been rejected in the leach residues, and therefore the solvent extraction process for sulphate liquors involves only steps (ii) to (v).[11,19]

The details of each part of the solvent extraction process are discussed below and a flowsheet is given in Fig. 1.

2.1.1   Iron removal. Iron is extracted from the pregnant chloride leach liquors, which have a pH of 1 to 2, with a 15 or 20 volume % solution of a secondary amine in kerosene, containing an equal volume % of isodecanol as a phase modifier. Two types of secondary amine are quoted in the patents: either N-lauryl-N-(1, 1-dimethyleicosyl)-amine (I), or N-lauryl-N-(1, 1-dimethylhexyl)-amine (II)

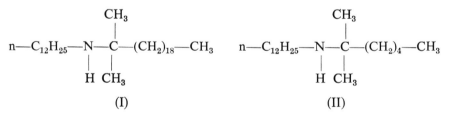

(I)   (II)

Extraction is carried out in four mixer-settlers. Stripping is done with dilute acid at pH 2 in three mixer-settlers to produce an aqueous solution of ferric chloride. This is then passed to a reactor, where water is evaporated and the $FeCl_3$ hydrolysed by reaction with water at a temperature of about 200°C to form HCl and Fe_2O_3. Alternatively the ferric chloride strip solution may be recycled to leaching and reacted with raw manganese nodules, whereby iron is converted to oxide and manganese is dissolved:[18]

$$4FeCl_3 + 3MnO_2 = 2Fe_2O_3 + 3MnCl_2 + 3Cl_2 \tag{1}$$

The iron oxide reports in the leach residues.

2.1.2   Copper removal. The iron-free chloride liquor or the pregnant sulphate liquor have the pH adjusted to approximately 2 by the addition of 2N NaOH solution. Copper is then selectively extracted with a solution containing 10 volume % LIX 64N, plus 20 volume % isodecanol in Napoleum, a paraffinic

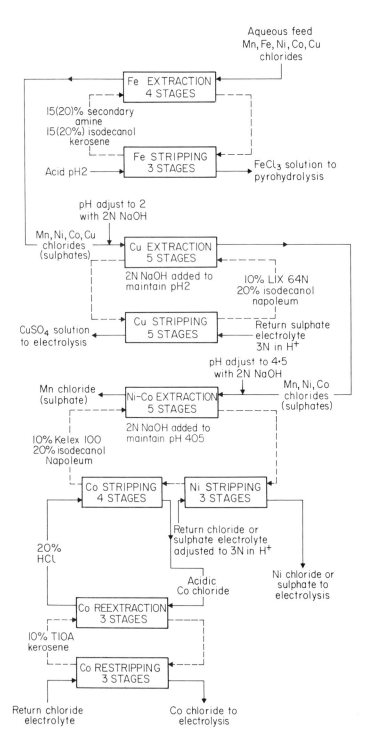

diluent. Five mixer-settlers are used for extraction and the pH is maintained at 2 in each stage by additions of NaOH. Copper is stripped from the loaded organic in five stages with sulphate electrolyte returned from the copper electrowinning step, which has a hydrogen ion concentration of at least 3N. In one patent, copper is extracted with an organic solution containing 10 volume % Kelex 100 instead of LIX 64N[15]. In this case, only 4 stages of extraction and 3 stages of stripping are used.

2.1.3   Nickel and cobalt removal. The raffinate from the copper removal step contains nickel, cobalt and manganese in the chloride or sulphate liquor. The pH of this liquor is adjusted to approximately 4·5 with 2N sodium hydroxide solution and then the nickel and cobalt are extracted together into an organic phase containing 10 volume % Kelex 100 and 20 volume % isodecanol in kerosene or Napoleum. Five stages of extraction are used and the pH is maintained at the desired value by additions of NaOH.

Nickel is then selectively stripped from the loaded organic phase, which contains nickel and cobalt, by contact in three stages with an aqueous solution 3N in hydrogen ions. This solution is return electrolyte from the nickel electrowinning step with added acid to increase the hydrogen ion concentration to the desired value. It is interesting to note that nickel electrowinning is done using a nickel chloride electrolyte when the liquor from nodule leaching is chloride based[15,18,20,21] and from a nickel sulphate electrolyte when the leach liquor is sulphate based.[11,19]

After nickel stripping, cobalt is stripped from the organic phase by contact in 4 stages with 20 weight % HCl solution. This produces an acidic cobalt chloride solution which is unsuitable for cobalt electrowinning and, in order to conserve the strong HCl solution for re-use in stripping, cobalt is re-extracted into a 10 volume % solution of tri-isooctylamine in kerosene or Napoleum. This extraction is carried out in three mixer-settlers. Stripping is carried out, also in three stages, with electrolyte returned from the cobalt electrowinning step.

In one patent, a 10 volume % solution of LIX 64N is used in place of Kelex for the co-extraction of nickel and cobalt[19]. In this case, only 3 stages of extraction are used but, apart from this, the process is identical to that described above.

The final aqueous raffinates from the solvent extraction processes are basically solutions of manganese chloride or manganese sulphate. These are usually further purified by precipitation of heavy metal impurities as sulphides, by the addition of hydrogen sulphide or ammonium sulphide,[18,19,21] resulting in substantially pure solutions of manganese chloride or sulphate.

Two alternative methods of producing manganese chloride solutions from pregnant chloride leach liquors are described in other patents. In both cases, iron

---

Fig. 1. Deepsea Ventures solvent extraction flowsheet. Adapted from US Patents 3 809 624; 3 894 139; 3 903 236; 3 923 615; 3 950 486.

is first removed from the leach liquors by amine extraction and then the remaining metals are removed together from solution by precipitation. This is accomplished either by cementation with manganese metal[12] or by the use of H_2S to produce a bulk precipitate of metal sulphides.[23] These precipitates will of course require further processing. The methods used are not revealed, but are likely to involve redissolution, followed by solvent extraction for metals separation.

2.2   Manganese recovery

The recovery of manganese metal from manganese sulphate solutions can be accomplished by aqueous electrowinning.[11,19] However, a problem arises in the disposal of the sulphuric acid generated in the electrowinning step. A partial solution to this problem is revealed in a patent which will be discussed in a later section.

The recovery of manganese from manganese chloride liquors requires novel technology and a number of possible solutions have been developed by Deepsea Ventures: recovery of manganese metal by either reduction with aluminium metal or sub-halide, or fused-salt electrolysis, or recovery of manganese as an oxide by pyrohydrolysis of $MnCl_2$.

2.2.1   Reduction with aluminium.

The operation of an aluminium reduction cell is outlined in US Pat 3 832 165.[12] $MnCl_2.4H_2O$ is crystallised from the aqueous liquor and dried. The crystals are added to a pool of molten chlorides contained in a refractory-lined reaction vessel. The pool contains about 50 % manganese chloride and the balance is alkali and alkaline earth chlorides. There is a vertical temperature gradient within the pool, with the top at 1150°C and the bottom at 1300°C. A substantially stoichiometric amount of scrap aluminium turnings is added together with the manganese chloride. Molten manganese metal is tapped from the bottom of the reaction vessel and aluminium chloride vapour is removed overhead through a vapour outlet.

The aluminium chloride vapour is passed through a tower containing the incoming aluminium scrap to preheat it and to recover any manganese chloride carried over. The tower is maintained at a temperature above the boiling point of aluminium chloride.

The aluminium chloride vapour is then passed into a reactor where it is contacted with water vapour at approximately 400°C. Aluminium oxide is formed and recovered as a by-product and HCl is collected and recycled to leaching.

2.2.2   Fused salt electrolysis.

An alternative method of producing molten manganese metal, which does not require a supply of scrap aluminium, is fused salt electrolysis.[13] Manganese halide is reduced in an electrolysis cell containing a molten mixture of halides. One component of the mixture is the halide of a reactant metal, preferably magnesium or aluminium. The electrodes in the cell are inert and the voltage drop is not sufficient to cathodically reduce the manganese halide to manganese metal. The process is believed to operate by reduction of the reactant metal halide to the elemental state. The reactant metal then reacts with the manganese halide to give manganese metal and to reform the reactant metal halide. Thus there is substantially no net loss of the reactant metal from the cell. Elemental halogen is formed and removed at the anode.

The reactor used is in two parts, with the upper portion being the electrolytic cell containing the electrodes. The current flow is horizontal. The lower portion is a molten manganese collection section. The molten halide bath in the cell section is a mixture of at least one alkali halide, at least one alkaline earth halide, the reactant metal halide and manganese halide. The alkali and alkaline earth halides that are quoted as useful are the chlorides, bromides and iodides of Na, K, Cs, Rb, Li, Ca, Ba, Sr and Mg. The halide mixture contains about 10 to 25 wt % reactant metal halide and up to 10 wt % manganese halide. In order to maintain the molten reactant metal as dispersed droplets in the halide mixture and thus to maximise the surface area for reaction with the manganese halide, an antiagglomerating agent such as B_2O_3, $Na_2B_4O_7$ or $K_2B_2O_4$ is present in the mixture at a concentration of 10^{-3} to 10^{-2} wt %. The manganese is collected in the molten state in the lower part of the reactor, which is maintained at a temperature of 1260–1300°C. In order to prevent loss of halides by volatilisation, there is a temperature gradient in the reactor and the top surface of the halide bath is kept in the range 800–1000°C.

2.2.3   Pyrohydrolysis. The production of manganese oxide rather than manganese metal is described in another patent.[23] The hydrated manganous chloride, $MnCl_2.4H_2O$, which is crystallised from the purified chloride leach liquors, is first dried to the anhydrous salt. This is passed to a pyrohydrolysis reactor where it is reacted with steam at 550°C to form manganese oxide and HCl, which can be recycled.

Manganese oxides can be converted to manganese metal by reaction with the sub-halides of one of the so-called transport metals, which comprise aluminium, silicon or titanium, according to US Pat 3 950 162.[34] The process, which is largely conceptual in design, consists of three reactors in closed circuit. Manganese oxide is charged to a reduction reactor where it is reacted with aluminium monohalide –

$$3MnO + 3AlCl \rightarrow 3Mn + Al_2O_3 + AlCl_3 \qquad (2)$$

The reaction is carried out at about 1350°C in the presence of calcium fluoride, which acts as a flux for the metal oxides. Molten manganese metal is tapped from the reduction reactor. Aluminium oxide, together with the flux, is removed as a slag and passed to a carbothermic smelter. Here the aluminium oxide is reduced with carbon at about 2100°C. Silicon is present, so that a molten aluminium-rich silicon alloy is formed. This prevents the formation of aluminium carbide :

$$Al_2O_3 + 3C + (Si) \rightarrow 2Al(Si) + 3CO \tag{3}$$

The molten alloy is then passed to the sub-halide reactor, where it is reacted at about 1150°C with the aluminium trichloride, returned from the reduction reactor, to reform aluminium monochloride :

$$2Al(Si) + AlCl_3 \rightarrow 3AlCl + (Si) \tag{4}$$

Similar principles can be used to reduce manganese chloride with aluminium monochloride to form manganese metal.

2.3   Front-end processes

A large proportion of DSV patents are concerned with alternatives for the front-end operations: i.e. the pretreatment and leaching of raw manganese nodules. As the objective is to dissolve manganese, it is necessary to reduce manganese from the tetravalent to the divalent state and this is done either prior to or during the leaching operations. Both chloride-based and sulphate-based routes have been investigated and it is convenient to discuss them on this basis.

2.3.1   Chloride-based processes.

The most direct method of producing a chloride liquor is dissolution of the nodules in concentrated hydrochloric acid. The acid is sufficiently reducing to reduce manganese to the soluble divalent state. The oxidation product of the reaction is elemental chlorine:

$$MnO_2 + 4HCl \rightarrow MnCl_2 + Cl_2 + 2H_2O \tag{5}$$

Direct leaching of manganese nodules in hydrochloric acid is quoted in at least four DSV patents. The nodules are ground to $-500~\mu m$ and then contacted counter-currently with 11M HCl at 100°C in a five-stage leaching system. The pregnant liquor is a mixture of $MnCl_2$, $FeCl_3$, $CoCl_2$, $NiCl_2$ and $CuCl_2$ with a pH of 1 to 2.[12,18,23] A modification of this process involves a sand-slimes split of the ground nodules in a hydroclassifier prior to leaching. The slimes are treated in a three-stage leaching system with counter-current flow of an 11M HCl solution. This solution is then contacted counter-currently with the sands fraction in a five-stage system. Additional HCl gas is added during the sands leaching.[15]

The great advantage of direct hydrochloric acid leaching is that the nodules, which contain about 30% moisture, do not have to be dried prior to leaching. However, as shown in the equation above, approximately half the hydrochloric acid is oxidised to chlorine during leaching. This chlorine must either be sold as a by-product or reconverted to HCl by reaction with hydrogen. Also most of the iron present in the nodules reports in the leach solution. An alternative way of using HCl involves chloridisation at temperatures around 500°C with gaseous HCl.[21] An advantage of this procedure is that by passing water vapour over the hot ore after chloridisation, iron can be converted to insoluble Fe_2O_3 and is thus eliminated from the subsequent leach liquor, produced by dissolving the chloridised ore in dilute acid at pH 2.

The dissolution of iron can also be prevented by chloridising the nodules in a molten chloride salt bath.[16,17] The nodules are ground to $-710 \mu m$, dried and mixed with a 48% NaCl, 52% $MgCl_2$ mixture. This is heated to 200°C for one hour and then to 600°C to form a liquid mass and held for six hours. The percentage conversion of the various metals to the chloride form is as follows: Mn – 35·5, Fe – 0·0, Ni – 76·5, Cu – 92·3, Co – 72·5. The metal chlorides are recovered from the molten chloride bath by raising the temperature to 1000°C to volatilise them. They are then carried out of the reactor in a stream of nitrogen gas and condensed in water. The extraction of Mn by this procedure can be improved to over 80% by pre-reducing the nodules with carbon monoxide at 600°C, prior to chloridisation.

The selective chloridisation of Mn, Ni, Co and Cu can be carried out at lower temperatures by using solid aluminium or ferric chlorides.[22] Again pre-reduction, with carbon for example, improves the extraction of manganese. The pre-reduced nodules are mixed with anhydrous $AlCl_3$ and heated at 140°C for 2 hrs. After cooling, the mixture is leached with dilute acid at pH 2, to give metal extractions of better than 95%.

2.3.2   Sulphate-based process. Sulphate leach liquors may be produced by using SO_2 as a reducing agent. Manganese can be selectively converted to $MnSO_4$ by reacting nodules ground to $-149 \mu m$ with SO_2 in the absence of oxygen in a fluidised bed reactor. The reaction is exothermic and the temperature rises to about 100°C during the reaction.[11] Leaching the reacted ore with water counter-currently in three stages produces a substantially pure solution of manganese sulphate from which manganese metal can be won by aqueous electrolysis. Ni, Co and Cu can be extracted from the solid residue by slurrying it in water and passing in air containing 10% SO_2. Under these conditions Ni, Co, Cu and any remaining Mn dissolve as sulphates, while iron remains insoluble. Alternatively, the selective sulphation of manganese is eliminated and the raw nodules are reacted directly with an SO_2/O_2 mixture to form the sulphates of Mn, Ni, Co and Cu, which can be water leached.[14]

An interesting variation, involving the use of sulphidic iron ores as reducing and sulphating agents, is revealed in US Pat 3 809 624.[10] A slight excess of sulphidic iron ore is mixed with the nodules and the two are ground together to 149–

710 μm. They are then roasted in an excess of air at 400–600°C to form iron oxide and metal sulphates. Leaching the hot ore with dilute acid at pH 2 results in a solution containing the sulphates of Mn, Ni, Co and Cu.

The aqueous electrolysis of manganese sulphate solutions produces a spent electrolyte which is basically dilute sulphuric acid. This cannot be recycled in the above sulphate-based processes and presents a disposal problem. An alternative approach, in which at least part of this sulphuric acid is recycled, is described in US Pat 3 923 615.[19] Raw nodules are ground to -297 μm and leached with recycled spent electrolyte containing 4 wt% H_2SO_4. Leaching is carried out for about 14 hours at 60°C, during which time most of the nickel and copper dissolves. Ferrous sulphate is then added to the leach liquor and leaching is continued for a further 6 hours. The ferrous sulphate acts as a reducing agent and solubilises the manganese and cobalt. Ferrous iron is oxidised to ferric oxide which reports in the residue. Air is bubbled through the solution during the final hour of leaching to ensure the total oxidation of iron. Metal extractions obtained by this process were Mn, 91%, Ni, 84%, Co, 87% and Cu, 81%.

3   Kennecott Copper Corporation processes

The distinguishing feature of all Kennecott's work is that leaching is carried out in ammoniacal solutions, the objective being to recover Ni, Cu, Co and Mo in solution and to reject Mn and Fe in the solid leach residues. The key unit operation of the various Kennecott processes is the solvent extraction separation of copper and nickel from the ammoniacal leach liquors. This is the subject of a number of patents and has been described in several papers.

The original solvent extraction scheme involved selective, sequential extraction of copper, followed by nickel, using LIX 64N as the extractant for both metals.[5,24] One of the main problems in using LIX 64N with ammoniacal aqueous solutions is that extraction of ammonia into the organic phase occurs. This has to be removed before the metals are stripped with sulphuric acid, otherwise ammonium sulphate builds up in the stripping/electrowinning circuits. Because of the necessity to include ammonia scrubbing stages in both the copper and the nickel circuits, a total of 26 stages were required for the selective extraction flowsheet. Detailed analysis of this flowsheet indicated that it could be considerably simplified by adopting co-extraction of Ni and Cu, followed by selective stripping,[24] and this flowsheet option has been the subject of subsequent papers from Kennecott.[25,26]

The flowsheet of the co-extraction/selective stripping process is shown in Fig. 2. The composition of the feed liquor is Ni, 6·2; Cu, 5·7; Co, 0·2; NH_3, 90; CO_2, 55 g/l. The extractant is 40% LIX 64N in kerosene and all extraction and stripping steps are carried out in mixer-settlers at a temperature of 40°C. Co-extraction of Ni and Cu into the organic phase requires 3 stages. Better than 99·9% extraction of both metals is achieved. About 5% of the ammonia is also extracted and the

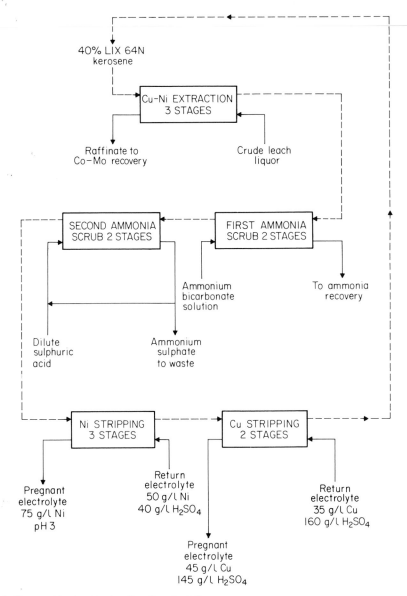

Fig. 2. Kennecott solvent extraction flowsheet.[24]

loaded organic phase contains 2.5 g/l NH_3. This is lowered to 0.005 g/l NH_3 in the two ammonia scrubbing sections. In the primary scrub section, ammonia is removed from the organic phase by contact in two stages with an ammonium bicarbonate solution containing 17 g/l NH_3. Ammonia is recovered from the aqueous scrub raffinate. Final traces of ammonia are removed in the secondary scrub section, by contact in two stages with an ammonium sulphate solution containing 27 g/l NH_3. The concentration of this aqueous scrub liquor is maintained by a bleed to waste. Nickel is then selectively stripped from the ammonia-free organic by contact with return electrolyte from the nickel electrowinning section. The return electrolyte contains H_2SO_4, 40; Ni, 50; Na_2SO_4, 100; H_3BO_3, 15 g/l. The advance electrolyte from nickel stripping contains 75 g/l Ni at pH3. The Ni/Cu ratio of this solution is 25000/1 and it is suitable as feed to nickel electrowinning without further treatment. The transfer of nickel from the organic to the aqueous phase is a slow process, requiring about 25 minutes contact time. In the pilot plant, this necessitated the use of six stages for nickel stripping to achieve the necessary contact time, but the proposed commercial design involves only three stages, where each stage consists of two mixers in series feeding a single settler.[26] The organic phase leaving nickel stripping contains 3·8 g/l Cu and 0·4 g/l Ni. Copper, together with the remaining nickel, is removed in the copper stripping section by contact in two stages with return electrolyte from copper electrowinning. The return electrolyte contains 160 g/l H_2SO_4 and 35 g/l Cu and the copper content is increased to 45 g/l in the copper stripping section. Nickel, which also transfers into the copper electrolyte, does not affect copper electrowinning provided its concentration is kept below 20 g/l. This is accomplished by a bleed from the copper tankhouse. The stripped organic, which contains about 0·5 g/l Cu, is recycled to extraction.

The development and optimisation of this flowsheet was considerably expedited by the use of a computer model developed by Kennecott.[25] The extraction system is chemically complex as there is competition between the metals and ammonia for the oxime reagent, ammine complexing in the aqueous phase and apparently the possibility of the extraction of a nickel ammine complex. The development of a successful computer model for such a complicated system was a considerable achievement. It was found to be particularly useful for this sytem because the extraction of nickel by LIX 64N is sharply decreased by increases in the aqueous ammonia concentration. On the other hand, leaching is improved by increases in ammonia concentration. Computer studies using the model enabled these two conflicting effects to be optimised.

In none of the published descriptions of this process, is the extraction of cobalt into the organic phase mentioned. We can assume, therefore, that cobalt in the feed liquor is in the cobaltic state, since cobaltic ammine complexes are not extractable by LIX 64N. Cobalt in the cobaltous state is extracted by LIX 64N. and in the organic phase it immediately oxidises and becomes difficult to strip. Kennecott hold two patents on the stripping of cobalt from LIX solutions. One method involves the use of a solution containing 2·5M HCl and 150–200 g/l NaCl. An aqueous to organic ratio of about 2 to 1 is used and the solutions are mixed for 30–60 minutes at 50–60°C.[27] Under these conditions about 90% of the cobalt is removed and the loss in metal loading capacity of the organic phase is held to < 3%. In a later patent,[28] mixtures of concentrated sulphuric acid with glacial acetic acid, methanol or other lower alcohols are cited. The acidic stripping mixture dissolves the cobalt-containing oxime, leaving the kerosene or other diluent as a separate phase. On entering the acidic mixture, the oxime releases the cobalt and

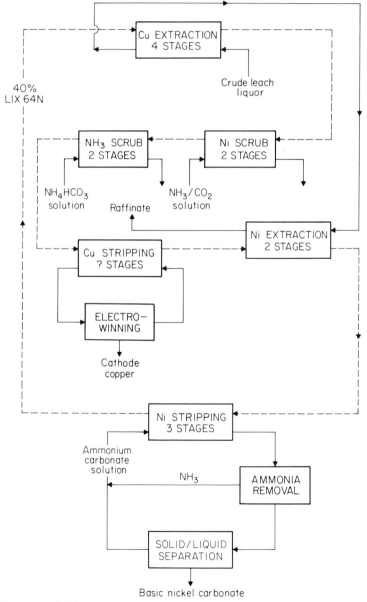

Fig. 3. Kennecott nickel carbonate process. Adapted from US Patent 3 907 966.

the oxime is recovered by adding water to the acidic mixture to reduce its solubility, whereupon the oxime redissolves in the diluent phase. Some degradation of the oxime occurs and the metal loading capacity is reduced by about 2·5%.

Although the co-extraction/selective stripping flowsheet described above appears to be the preferred option, an alternative solvent extraction process involving selective, sequential extraction of copper and nickel is revealed in two patents.[29,32] The objective of this process is to recover nickel, as metal or oxide, without having to use electrolytic methods. This is accomplished by stripping copper-free, nickel-loaded LIX 64N with a concentrated ammoniacal ammonium carbonate solution. Removal of ammonia from the aqueous strip liquor precipitates basic nickel carbonate. A slurry of the basic nickel carbonate can be hydrogen pressure reduced to produce nickel metal powder. Alternatively the basic carbonate can be calcined to produce nickel oxide. This can be marketed as such, or reduced at high temperatures to nickel metal powder. The solvent extraction flowsheet for this process is shown in Fig. 3.

Kennecott have not published details of the leaching process to be used to produce the ammoniacal feed solutions for their solvent extraction processes. However the patent literature indicates that the preferred process is direct leaching of ground raw nodules using ammoniacal ammonium carbonate solutions containing cuprous ions. The cuprous ions, which are stabilised in solution by ammine complexing, act as reducing agent and reduce manganese dioxide in the nodules, releasing Ni, Cu, Co and Mo into solution, while the manganous ions precipitate as insoluble manganous carbonate. The cuprous ions are continuously regenerated during leaching by introducing carbon monoxide into the leaching reactors.[31]

The reactions which occur are said to be:

$$MnO_2 + 2Cu(NH_3)_2{}^+ + 4NH_3 + CO_2 + H_2O \rightarrow MnCO_3 + 2Cu(NH_3)_4{}^{2+} + 2OH^- \tag{6}$$

$$2Cu(NH_3)_4{}^{2+} + CO + 2OH^- \rightarrow 2Cu(NH_3)_2{}^+ + 4NH_3 + CO_2 + H_2O \tag{7}$$

The net overall reaction for the reduction is the sum of equations (6) and (7):

$$MnO_2 + CO \rightarrow MnCO_3 \tag{8}$$

In the absence of copper in solution, reaction (8) does not occur and thus copper can be considered as acting as a catalyst for the reduction of manganese dioxide by carbon monoxide.

A description of a flowsheet for atmospheric pressure leaching is given in US Pat 3 983 017[31] and is shown in simplified form in Fig. 4. Raw nodules are ground

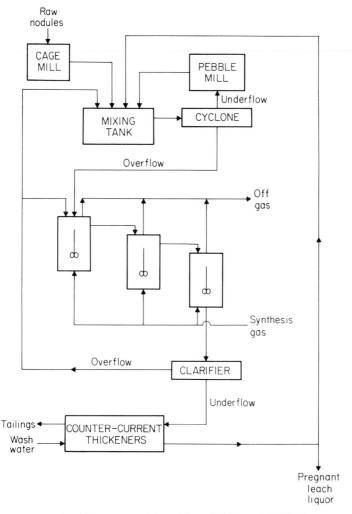

Fig. 4. Kennecott cuprous leaching process. Adapted from US Patent 3 983 017.

to $-3360\ \mu m$ in a cage mill and fed to a mixing tank where they are slurried with pregnant liquor recycle and clarified leach liquor. The slurry is passed through a hydrocyclone, where oversized nodule particles are removed, reduced in size in a pebble mill and returned to the mixing tank. The hydrocyclone overflow, together with part of the clarified leach liquor is fed to the first of the impeller-agitated leaching tanks. Co-current leaching at ambient temperature and pressure and about 5% solids, is carried out in a series of from three to six tanks. The composition of the leach liquor is specified in the patent to be within the following broad limits: about 60–140 g/l NH_3, about 20–60 g/l CO_2, about 2–20 g/l or more copper, about 8–20 g/l Ni, about 0·5–4 g/l Co and about 10–20 g/l Cl. The chloride arises presumably from the sea water contained in the raw, undried nodules. The necessary cuprous ion concentration in the leach liquor is maintained by the introduction of synthesis gas into each leaching reactor. The synthesis gas contains 40–60% CO, 30–45% H_2, 6–12% H_2O and about 1% N_2. The off-gas from the leaching reactors consists primarily of hydrogen, with some ammonia and unreacted carbon monoxide. Slurry from the last leach tank is passed to a clarifier, where it is thickened to 40–50% solids. The thickened slurry is then washed counter-currently with an aqueous ammoniacal ammonium carbonate solution. Washing is carried out in a series of from 3 to 8 thickeners and produces a solid tailings for manganese carbonate recovery or disposal and a pregnant liquor containing nickel, copper, cobalt and molybdenum. A portion of the pregnant liquor is recycled to the mixing tank and the remainder goes forward for metals separation and recovery. The overflow from the clarifier is cooled in a heat exchanger and then part is recycled to the first leaching reactor and the rest is returned to the mixing tank.

Improvements on this process, involving the use of pressurised leaching reactors, are revealed in further patents.[33] Comminuted raw nodules are leached at 35–55°C with ammoniacal ammonium carbonate solutions in which the cuprous ion concentration is maintained at greater than 2 g/l by using carbon monoxide at pressures of 3·5 to 7 kg/cm^2 (50 to 100 psi). The gas is introduced concurrently with the slurry flow in the leaching train, while the manganese nodules are injected into several leaching reactors simultaneously. The heat of reaction is removed by inter-stage heat exchangers so that each leaching reactor is operated at substantially the same temperature. Under these conditions, the efficiency of the leaching process is said to be greatly improved, enabling the size of the leaching reactors to be reduced.

The possibility of recovering copper and nickel by solvent extraction directly from the leach slurries produced by cuprous ion leaching, without prior thickening and filtration, is described in other patents.[30] The slurry from leaching is diluted with ammonium carbonate solution to obtain a slurry of about 17 to 20 wt % solids, with a pH of 9·5. This is fed to a mixer-settler cascade where Ni and Cu are co-extracted at 40°C with LIX 64N solutions. Low-speed mixing is used to maintain an organic-continuous, uniform suspension, with slurry droplet sizes of the order of 1 mm diameter. Under these conditions, it was found that satisfactory metals extraction could be achieved, while holding losses of LIX 64N, by adsorption and entrainment in the slurry, to within 100 to 200 ppm. Proper control of mixer operation, solids content of the slurry and especially pH, were important in minimising reagent loss. For example, at pH 11, the loss of reagent was almost 2 g/l of slurry.

4   Other companies

4.1   *International Nickel Company (INCO)*

INCO's approach to nodule processing differs from the others described in this review in that the initial separation of manganese and iron from the other metals is carried out pyrometallurgically. A combination of roasting and smelting is used to produce a slag containing most of the manganese and a matte containing Ni, Cu and Co, together with other impurities. The matte is then treated hydrometallurgically to recover the valuable metals, while the slag can either be discarded, or converted to marketable ferromanganese by reduction smelting. The process has been described in a paper by Sridhar, Jones and Warner.[2]

The first step of the process is selective reduction of the raw nodules in a rotary kiln. The nodules are dried and pre-heated in the first part of the kiln and are then reduced at 1000°C to convert most of the nickel, copper and cobalt and part of the iron to the metallic state. The necessary reducing atmosphere in the kiln is produced by combustion of fuel oil with sub-stoichiometric amounts of air and by the separate addition of reductant, such as coal, to the reduction zone. Total residence time in the kiln is approximately 2 hours. The hot reduced nodules are then passed to an electric furnace for smelting to produce a fluid slag containing the manganese and a molten alloy containing the reduced metals. Smelting is carried out for one hour at 1380–1420°C under a reducing atmosphere and fluxes are added if necessary. Simulated tests of the reduction and smelting steps have shown the following recoveries of metals into the alloy phase: Ni, 93–98%; Cu, 85–95%; Co, 90–98%; Fe, 80–90%; Mn, 0·5–2·5%. The alloy phase, which is 6–8·5% by weight of the original nodule feed, also contains most of the Mo, As, Sb and Zn present in the feed. The next step is the production of a matte by sulphiding the alloy with elemental sulphur. In order to achieve high sulphur efficiency, it is necessary to lower the manganese content of the alloy to < 0·1%. This is done by prior oxidation of the alloy with air to oxidise the manganese, which is removed as a fluid slag by fluxing with silica. About 10 to 15% of the iron also enters this slag. Elemental sulphur is then added to form a matte, which is then converted to eliminate the remainder of the iron. Mattes produced in this way contain typically, Cu, 25; Ni, 40; Co, 5; Fe, 5; S, 20–25%, together with minor amounts of Sb, P, As and Mn.

Marketable ferromanganese can be produced from the slag obtained in the smelting operation. The major problem is meeting the phosphorus specification for ferromanganese, as the Mn/P ratio in the nodules is approximately 100, whereas it has to be better than 250 in ferromanganese. The distribution of phosphorus, and also of manganese, between the alloy phase and the slag depends on the extent of reduction of the nodules. It has been established that controlling reduction to yield alloys containing $\leqslant 1·5\%$ Mn produces slags with acceptable Mn/P ratios. Ferromanganese can be produced from these slags by reduction smelting with added lime at approximately 1600°C.

The valuable metals are recovered from the matte by hydrometallurgical methods. The matte is ground to $-44\ \mu m$ and then oxygen-pressure leached with sulphuric acid. The preferred leaching conditions are 100 g/l H_2SO_4, 9% pulp density and an oxygen partial pressure of 1·0MPa ($\sim$ 150 psi). Leaching is carried out at 110°C with a 2 hour residence time. Under these conditions, metal extractions of 99% are achieved and 99% of the sulphide sulphur is oxidised to sulphate. It appears that these conditions are preferred in order to minimise the amount of leach residue, which is only about 1% by weight of the matte feed. However it would appear that by choosing to oxidise sulphur in the matte to sulphate, rather than elemental sulphur, a problem is being created in the eventual disposal of this sulphate.

The liquor from leaching typically contains Ni, 40; Cu, 24; Co, 5; Fe, 5; free H_2SO_4, 20 g/l. Details of the separation procedures to be used are not given, except that iron is removed by oxidation and precipitation with limestone at a pH of 3·5. This has the advantage of removing many of the impurities by co-precipitation. The impurities left in the iron-free solution are Mn, Bi, Se and Zn, all at concentrations less than 10 ppm. It is suggested that a possible separation scheme would involve solvent extraction with a LIX reagent to remove copper, and either solvent extraction or selective oxidation and precipitation to remove cobalt, leaving a pure nickel solution.

4.2   Metallurgie Hoboken-Overpelt

In common with much of the work carried out by Deepsea Ventures, the process developed by Metallurgie Hoboken-Overpelt (MHO) uses strong hydrochloric acid as the leaching agent for manganese nodules.[35] In contrast with Deepsea Ventures, however, MHO have an elegant solution to the disposal of chlorine generated by this method of leaching. Chlorine from the leaching reactors is passed to the end of the process, where it is used to oxidise the manganous chloride end-solution. Under conditions of controlled pH, obtained by neutralisation with MgO, manganese is thus precipitated from solution as mixed manganese oxides, while chlorine is reduced back to chloride. This then enables the regeneration of hydrochloric acid from the demanganised solution for recycle to leaching. The purification of the crude chloride leach liquor is also quite different from that proposed by Deepsea Ventures, involving not only solvent extraction, but also a number of precipitation steps. The MHO process is designed to effect separations of all the valuable metals contained in the manganese nodules. A flowsheet of the process is given in Fig. 5.

Leaching of raw nodules, ground to less than 2 mm diameter, is carried out with hydrochloric acid in up to six stages. Most of the acid required for leaching is obtained by scrubbing the exit gases issuing from the final step of the process, which is pyrohydrolysis of the demanganised solution. Gas scrubbing is carried out ahead of each of the leaching stages. In the first stage, scrubbing is done with water, but in subsequent stages, the leach pulp itself is used to scrub the pyrohydrolysis gases.

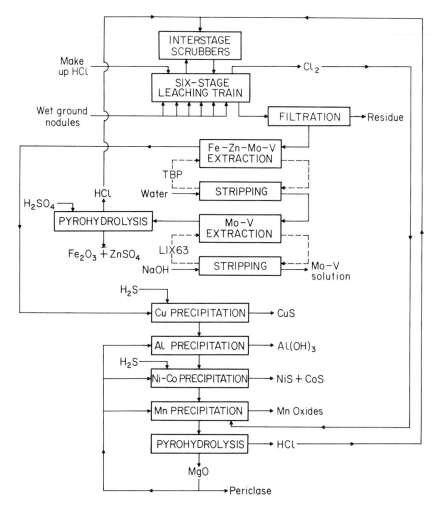

Fig. 5. Metallurgie Hoboken-overpelt process. Adapted from US Patent 4 026 773.

The initial acidity in the first leach reactor is in the range 250–350 g/l HCl, obtained by adding make-up HCl to the scrub liquor. In the other leaching re-actors, the initial acidity is about 200 g/l HCl. Ground nodules are added to each reactor such that the initial HCl content in that reactor is 5 to 20 % higher than the stoichiometric amount required to dissolve the total metal content of the nodules as chlorides. Leaching is carried out at 70°C, this temperature being maintained by the exothermic nature of the reactions, and the residence time in each stage is 30 minutes. The final leach liquor contains between about 110 and 120 g/l Mn and the following metal extractions are claimed: Mn, 99·9 %; Ni, 99·9 %; Co, 99·5 %; Cu, 99·9 %. Typically the final leach liquor will contain, in addition to manganese, the following approximate concentrations of other metals: Cu, 5; Ni, 5; Co, 1; Zn, 0·6; V, 0·3; Mo, 0·2; Mg, 7; Fe, 25 g/l.

The first stage of purification of this liquor involves the use of undiluted tributyl phosphate (TBP) to co-extract Fe, Mo, V and Zn. This is done in 3 stages of extraction with an organic/aqueous ratio of 1 : 2. Any trivalent vanadium in the leach liquor has to be oxidised to the pentavalent state prior to TBP extraction. Better than 99·9% extraction of all four metals is achieved in this step. Water is used to strip the loaded TBP. The aqueous strip liquor, containing Fe, Mo, V and Zn, is then contacted with LIX 63, which extracts Mo and V. These metals are stripped with dilute NaOH to produce a mixed aqueous sodium molybdate and vanadate solution. Sulphuric acid is added to the aqueous raffinate from the LIX 63 extraction, which contains the iron and zinc as chlorides, and then this solution is pyrohydrolysed in a spray-roaster to produce HCl for recycling and an Fe_2O_3–$ZnSO_4$ mixture. Water leaching of the latter material produces a zinc sulphate solution and an iron oxide residue.

The main leach liquor coming from the TBP extraction stage is next treated to remove copper. This is done by selective precipitation of copper sulphide using hydrogen sulphide gas, with the pH maintained below 2·5. The reaction is carried out at 20°C, with the H_2S being injected counter-currently to the solution in a column. The feed rate is controlled such that the H_2S is completely consumed at the top of the column.

Aluminium is then precipitated as the hydroxide by neutralisation of the solution with magnesia to a pH of between 2·5 and 5. Finally, nickel and cobalt are co-precipitated from solution as sulphides, again using H_2S, in a similar manner to copper, except that the pH is maintained at about 4 by continuous neutralisation with magnesia. The mixed nickel-cobalt sulphides can then be further processed by oxygen-pressure leaching with hydrochloric acid to form a nickel-cobalt chloride solution and elemental sulphur. The latter is removed by filtration and any sulphate formed during the oxidative leaching is removed by the addition of calcium or barium salts to the solution. The nickel-cobalt separation is then achieved by solvent extraction of cobalt from the chloride liquor with a tertiary amine.

The main leach liquor, which now contains substantially only manganese and magnesium, is treated to remove manganese from solution. As already indicated, chlorine generated in the leaching reactors is used to oxidise and precipitate the manganese. At first sight, it may seem a thermodynamic impossibility to first use manganese dioxide to oxidise HCl to chlorine and then to use chlorine to oxidise a manganese solution to precipitate oxides. The key to this conundrum is the acidity of the solution. The equilibrium involved can be represented, in simplified form as:

$$Mn^{2+} + Cl_2 + 2H_2O \rightleftharpoons MnO_2 + 4H^+ + 2Cl^- \qquad (9)$$

Under conditions of high acidity, as used in leaching, the equilibrium of the above reaction lies well to the left, and manganese dioxide oxidises chloride to chlorine and is itself reduced to the soluble manganous state. However, at lower acidities, in the pH range 0·5 to 3, the equilibrium is reversed and chlorine can be used to oxidise and hydrolyse manganous ions to manganese dioxide. If the pH is raised further into the 3·5 to 5·5 range, then Mn_2O_3 becomes the stable oxide species:

$$2Mn^{2+} + Cl_2 + 3H_2O \rightleftharpoons Mn_2O_3 + 6H^+ + 2Cl^- \qquad (10)$$

It may be seen that the precipitation of manganese as Mn_2O_3 consumes only half the chlorine required for the formation of MnO_2. The MHO process takes advantage of this fact and, by pH control, manganese is precipitated substantially as Mn_2O_3, or its hydrated form, $MnOOH$. Precipitation is carried out by passing a measured amount of chlorine into the manganous chloride solution contained in a stirred reaction vessel. Magnesia is added continuously to neutralise the acid produced by the hydrolysis reaction and the pH is maintained in the range 3·5 to 4. Virtually total precipitation of manganese occurs within one hour, from solutions containing typically 130–140 g/l Mn.

The final demanganised solution is basically a pure magnesium chloride solution, containing about 120 g/l Mg. This is pyrohydrolysed in a spray-roaster to form HCl for recycle to leaching and MgO, part of which is recycled to the various neutralisation steps and the remainder is converted to periclase.

Thus the cycle of this elegant process is almost totally closed. In principle, the only reagents consumed are the H_2S used to precipitate Cu, Ni and Co and the make-up HCl added to the first leaching reactor. All other reagents are internally regenerated and all the metals of any significance in the nodule feed leave the process in usable or disposable forms.

5   Conclusions

Any attempt to compare and draw conclusions about the technical merits of metallurgical processes, using information drawn mainly from patents, is a task fraught with difficulties. By their very nature, process patents are not very informative documents, particularly if the patent agent has done his job properly! Technical problems are generally glossed over and, until a process is in operation, at least at the pilot-plant stage, and the developers are prepared to publish frank and detailed technical papers, a realistic assessment is very difficult.

Nevertheless it is possible to make certain broad comparisons between the approaches taken by the various companies to manganese nodule processing. For example, there is little doubt that Kennecott have made a very realistic and pragmatic approach to the problem, and their process has many attractive features. In contrast, the sheer numbers and diverse natures of the patents assigned to Deepsea Ventures strongly suggest that many of the patented ideas are not technically satisfactory and that there has been a continuous search for new routes. This difference is perhaps not surprising in view of the fact that Kennecott are an established metallurgical company with an enormous experience in metals production, whereas Deepsea Ventures are virtually unknown in the metallurgical world.

The pragmatism of Kennecott's approach is evident in a number of ways. The primary decision, to base the economics of the overall process on the recovery of nickel, cobalt, copper and molybdenum only and to discard manganese, is perhaps the most important evidence of hard commercial realism. It is beyond the scope of this review to enter into the economic aspects of manganese nodule processing, but suffice it to say that any process which depends for its economic viability on the sale of manganese, as well as the other metals, must be considered a doubtful commercial proposition. This problem has been discussed at length in other papers.[5-7] Briefly it is related to the scale of production of manganese from nodules, compared with the present size of the world market for manganese. The perturbation that would be caused by the former on the latter means that the price that could be obtained for nodule manganese is a matter of considerable speculation.

The ammoniacal route chosen by Kennecott has a number of technical advantages. The chemistry of ammoniacal ammonium carbonate solutions is reasonably well understood and the reagents are easily recyclable. Neither iron nor manganese is soluble in ammonium carbonate solutions and so the primary separation of these major components from nickel, copper, cobalt and molybdenum is achieved at the leaching stage. The latter metals are readily separated by solvent extraction and the buffering effect of the ammoniacal ammonium carbonate medium eliminates any necessity for pH adjustment during extraction. The use of LIX reagents, which are cation exchangers, to extract nickel and copper enables the transfer of these metals to be made into sulphate media, by stripping with sulphuric acid. The metals can thus be recovered using the standard technology of electrowinning from sulphate solutions.

In the past, the major disadvantage of ammoniacal routes for nodule processing was the necessity to reduce the contained metals to the elemental state by reduction roasting prior to leaching. This is, of course, a very energy intensive step, due to the high moisture content of raw nodules. The new route devised by Kennecott, where direct ammoniacal leaching of raw nodules is accomplished by utilising the reducing properties of carbon monoxide, with dissolved copper acting as a catalyst, appears to be a very innovative and elegant solution to the problem. Not only does this procedure eliminate the roasting step, but it does so without the introduction of any extraneous chemical species into the leach solutions.

While some of the chemistry of the Kennecott process is innovative, the process technology involved is tried and tested. Design and scale-up of the unit operations should therefore be relatively straightforward. Corrosion problems in ammoniacal systems are minimal and standard materials of construction can be used. The process has been run at the pilot-scale and there has been a thorough study of the behaviour of trace impurities in the system. There is thus little doubt that Kennecott are in a position to build a full-scale nodule processing facility whenever this should be required.

The position of Deepsea Ventures however seems very different. There is much less hard information available about their intended process route. The diverse nature of the many patents makes identification of the preferred options very difficult. It is known that a chloride route has been investigated at the pilot-scale,[3] but there are few details available. It seems certain that the process requires that manganese be produced, if only to recover the associated chloride for recycle. While technically possible, this step will require the use of untried technology, and furthermore the economic uncertainties involved in the production of manganese from nodules have already been referred to.

The two unit operations in the chloride route which can be identified from the Deepsea Ventures patent literature, namely the leaching and solvent extraction steps, both appear to have technical drawbacks. In particular, the solvent extraction process has at least two serious disadvantages. These both stem from the decision to use chelating reagents, namely LIX or KELEX, to selectively extract first copper and then nickel and cobalt. Continuous neutralisation with NaOH is necessary to do this. Not only is this neutralisation agent not recyclable but also chloride will be lost from the system in the form of sodium chloride. Stripping cobalt from chelating reagents is extremely difficult, because cobalt oxidises to the trivalent state in the organic phase. While this allows a good separation of nickel from cobalt by selective stripping, the subsequent stripping of cobalt requires very rigorous conditions. Deepsea Ventures use 20 wt % HCl to strip cobalt and there is little doubt that serious degradation of the extractant will occur in the long term under these conditions.

The major disadvantage of the Deepsea Ventures leach system, where strong HCl is used to dissolve raw nodules, with the result that approximately half the acid is oxidised to chlorine during the leach, is the fact that this chlorine is not utilised in the remainder of the process. It thus has either to be sold as a by-product or converted back to hydrochloric acid. In the latter case, a separate hydrogen plant would be required to supply the hydrogen necessary for acid production.

Most of the disadvantages of the Deepsea Ventures chloride routes have been avoided in the elegant chloride process devised by Metallurgie Hoboken-Overpelt. The very ingenious use of the potential-pH relationships in the manganese-chlorine-water system, enables chlorine, generated in the leaching step, to be re-cycled within the process and converted back to HCl during the oxidation and precipitation of manganese oxides from the end-solution. Furthermore, the judicious combination of solvent extraction and precipitation steps used by MHO for solution purification and metals separation avoids the drawbacks of the Deepsea Ventures solvent extraction method outlined above. Finally the choice of magnesia as the only neutralisation agent in the process, allows recovery and recycle of both acid and neutralising agent, because $MgCl_2$ can be readily pyrohydrolysed to form MgO and HCl. Thus the MHO process is almost completely closed-cycle, an important attribute of any new process, which will have to comply with modern environmental legislation. The MHO process bears the hall-mark of the company, which has a long history of specialisation in the treatment and maximum utilisation of complex metal-bearing materials. The company's vast experience has obviously been used to good effect to produce a very attractive solution to the new extractive metallurgical problems posed by deep-sea manganese nodules.

In contrast with the three companies mentioned above, which all use direct leaching of raw nodules, INCO has opted for a smelting route. This has two main advantages. Firstly, virtually all the manganese and much of the iron is rejected into an inert slag, which can be readily discarded or possibly utilised to produce ferromanganese. Secondly, the valuable metals are greatly concentrated into an alloy phase, which is less than 10 wt % of the feed material. The alloy is converted to a matte and then treated hydrometallurgically, and, while few details of the process to be used are available, obviously the leaching and solution purification operations will be considerably smaller in volume than those required for direct leach processes. However any advantage in reduced capital expenditure gained at this end of the process will be offset by the cost of the reduction kiln and electric furnace required for smelting. Furthermore the requirement to dry and then melt large quantities of material with a high moisture content, makes energy a very heavy burden on the operating costs of any smelting route. There is no doubt that the weight of metallurgical opinion is in favour of direct hydrometallurgical processing of raw manganese nodules.

6   References

1 Mero J.L. Ocean Floor Manganese Nodules. *Econ. Geol.*, 1962, **57**, 747–67.
2 Sridhar R., Jones W.E. and Warner J.S. Extraction of copper, nickel and cobalt from sea nodules. *J. Metals*, April 1976, 32–37.
3 Cardwell P.H. Extractive Metallurgy of Ocean Nodules. *Min. Cong. J.*, Nov. 1973, 38–43.
4 Hubred G. Deep-Sea Manganese Nodules: A Review of the Literature. *Minerals Sci. Engng.*, 1975, **7**(1), 71–85.
5 Agarwal J.C. *et al.* Processing of Ocean Nodules: A Technical and Economic Review. *J. Metals*, April 1976, 24–31.
6 Moncrieff A.G. and Smale-Adams K.B. The Economics of First Generation Manganese Nodule Operations. *Min. Cong. J.*, Dec. 1974, 46–52.
7 Tinsley C.R. Economics of Deep Ocean Resources – A Question of Manganese or No-Manganese. *Min. Engng.*, April 1975, 31–35.
8 Flett D.S. and Spink D.R. Solvent Extraction of non-ferrous metals: A review 1972–1974. Hydrometallurgy, 1976, **1**, 207–240.
9 Flett D.S. Solvent Extraction of non-ferrous metals: A review 1975–1976. Rep. No. LR 254(ME), 1977, Warren Spring Laboratory, Stevenage, U.K.

10 Kane W.S. and Cardwell P.H. Mixed ore treatment of ocean floor nodule ore and iron sulphide land-based ores. U.S. Patent 3 809 624, May 7, 1974.

11 Kane W.S. and Cardwell P.H. Method for separating metal values from ocean floor nodule ore. U.S. Patent 3 810 827, May 14, 1974.

12 Kane W.S. and Cardwell P.H. Process for recovering manganese from its ore. U.S. Patent 3 832 165, Aug. 27, 1974.

13 Barton B.E. and Cardwell P.H. Fused salt electrolysis to obtain manganese metal. U.S. Patent 3 832 295, Aug. 27, 1974.

14 Kane W.S. and Cardwell P.H. Reduction method for separating metal values from ocean floor nodule ore. U.S. Patent 3 869 360, Mar. 4, 1975.

15 Cardwell P.H., Kane W.S. and Olander J.A. Nickel-cobalt separation from aqueous solution. U.S. Patent 3 894 139, July 8, 1975.

16 Kane W.S., McCutchen H.L. and Cardwell P.H. Recovery of metal values from ocean floor nodules by halidation in molten salt baths. U.S. Patent 3 894 927, July 15, 1975.

17 Kane W.S. and Cardwell P.H. Method of ocean floor nodule treatment and electrolytic recovery of metals. U.S. Patent 3 901 775, Aug. 26, 1975.

18 McCutchen H.L., Kane W.S. and Cardwell P.H. Method for obtaining metal values by the halidation of a basic manganiferous ore with ferric chloride pretreatment. U.S. Patent 3 903 236, Sept. 2, 1975.

19 Kane W.S. and Cardwell P.H. Winning of metal values from ore using recycled acid leaching agent. U.S. Patent 3 923 615, Dec. 2, 1975.

20 Cardwell P.H., Kane W.S. and Olander J.A. Separations of metals leached from ocean floor nodules. U.S. Patent 3 903 235, Sept. 2, 1975.

21 Cardwell P.H. and Kane W.S. Method for separating metal constituents from ocean floor nodules. U.S. Patent 3 950 486, Apr. 13, 1976.

22 Sandberg R.G. and Cardwell P.H., Halidation of nonferrous metal values in manganese oxide ores. U.S. Patent 3 990 891, Nov. 9, 1976.

23 Cardwell P.H. and Kane W.S. Halidation of manganiferous ore to obtain metal values and recycle of halide values. U.S. Patent 3 992 507, Nov. 16, 1976.

24 Agarwal J.C. et al. A new FIX on metal recovery from sea nodules. Eng. Min. J., Dec. 1976, 74–78.

25 Brown C.G. et al. Modelling a Fluid Ion Exchange System. Paper presented to ISEC 77, Int. Solv. Extn. Conf., Toronto, 1977.

26 Agarwal J.C. and Klumpar I.V. The role of solvent extraction and ion exchange in the processing of complex solutions. Paper presented to Joint IMM/SCI Conf. – Impact of SX and IX on Hydrometallurgy, Salford, 1978.

27 Skarbo R.R. Cobalt stripping from oxime solutions. U.S. Patent 3 849 534, Nov. 19, 1974.

28 Skarbo R.R., Galin W.E. and Natwig D.L. Cobalt stripping from oximes. U.S. Patent 3 867 506, Feb. 18, 1975.

29 Skarbo R.R. Nickel extraction and stripping using oximes and ammoniacal carbonate solution. U.S. Patent 3 907 966, Sept. 26, 1975.

30 Pemsler J.P. and Litchfield J.K. Solvent-in-pulp extraction of copper and nickel from ammoniacal leach slurries. U.S. Patent 3 950 487, Apr. 3, 1976, corresp. German Offen 2 526 395, Jan. 2, 1976.

31 Szabo L.J. Recovery of metal values from manganese deep sea nodules using ammoniacal cuprous leach solutions. U.S. Patent 3 983 017, Sept. 28, 1976.

32 Skarbo R.R. and Natwig D.L. Liquid ion exchange process for recovery of copper and nickel. U.S. Patent 3 988 151, Oct. 26, 1976.

33 Barner H.E., Kust R.N. and Cox R.P. Elevated pressure operation in the cuprion process. U.S. Patent 3 988 416, Oct. 26, 1976, corresp. German Offen 2 520 388, Jan. 2, 1976.

34 Schobert H.H., Field R.C. and Cardwell P.H. Reduction to manganese metal using metal transporting compounds. U.S. Patent 3 950 162, Apr. 13, 1976.

35 Van Peteghem A.L. Extracting metal values from manganiferous ocean nodules. U.S. Patent 4 026 773, May 31, 1977.

Extractive metallurgy of uranium

A. R. Burkin

1   The Uranium extraction industry before 1976

Until the early 1940s pitchblende deposits were worked for their radium content and the major constituent, uranium, was discarded because there was no forseeable future for it at that time. By 1942 work on the development of nuclear weapons led to the recognition of the strategic importance of uranium and a search for sources of supply by the British and United States governments.

Small quantities of radium had been produced in Australia between 1906 and 1931 and exploration for uranium began in earnest in 1944, leading to the discovery and exploitation of the Rum Jungle, South Alligator Valley, and Mary Kathleen deposits. In South Africa sufficient uranium was found to be present in some of the gold ores of the Witwatersrand to make its extraction from the material on the tailings dumps and from currently mined ores worth while. The early work on the development of processes to recover the 0·03 % of uranium contained in the ore was described in two volumes.[1] In Canada the pitchblende mine on the shore of Great Bear Lake in the far north was re-opened and in 1944 all the shares of the owners, the Eldorado Gold Mining Company were acquired by the federal government and a new Crown Corporation was formed, Eldorado Mining and Refining Limited. Uranium was also found in considerable quantities in the United States of America, making that country potentially independent of outside sources of supply.

In the early 1950s exploration for uranium increased dramatically because of the programmes in the United States and in Britain to construct thermonuclear weapons. In order to provide an incentive to prove substantial reserves of uranium special prices were offered, calculated on a cost plus basis and in the region of $10 per pound U_3O_8 in 1953-4. By 1956 more than 10 000 radioactive occurrences had been found in Canada alone and it was recognised that uranium was no longer in short supply. As new plant was brought into production output continued to rise until in May 1958 the government of Canada announced that in future private producers would be permitted to make their own arrangements for the sale of any surplus uranium instead of selling it to Eldorado Mining and Refining. In 1959 Canadian production reached a peak total of almost 16 000 tons of U_3O_8, having a value on the export market exceeding $330M, the highest for any mineral or metal exported from Canada that year. In November 1959 the United States Atomic Energy Commission announced that it would not take up it options to purchase additional uranium from Canada and in consequence the scale of operations had to be substantially reduced.

In order to prevent a total collapse of the uranium production industry by 1962 or 1963, when existing contracts were due to expire, the Canadian government negotiated longer delivery periods for uranium already under contract and initiated a stockpiling programme, for social rather than economic reasons. The first stockpile, accumulated over twelve months in 1963-4, consisted of 2700 tons of U_3O_8, the total expenditure being $24·5M. A second stockpiling programme between July 1965 and July 1970 led to the purchase of about 6950 tons of U_3O_8 at a cost of $76·9M. The average price paid for the two stockpiles was, therefore, $5·25 per pound U_3O_8.

In July 1970 the Canadian government announced that it was prepared to consider a programme of assistance for the uranium industry and was particularly concerned to ensure that the community of Elliot Lake be protected from a shutdown by established producers there. Discussions with Denison Mines Limited led to a joint stockpiling venture between the government and the company, 1000 tons of U_3O_8 to be acquired in each of the years 1971, 1972 and 1973. Early in 1972 it was agreed that the whole of the Canada-Denison stockpile would be committed in a single sale to several electricity utilities in Spain. In fact the quantity purchased by the Spanish exceeded the amount in the joint stockpile by about 1200 tons and this amount was provided from the government general stockpile. A further 1,000 tons of U_3O_8 was sold from the general stockpile to a Japanese electricity utility so that the joint stockpile was fully committed before it was accumulated and the general stockpile was reduced to 7500 tons.

2   The uranium market in the early 1970s

In 1971 and 1972 the international uranium market was in a chaotic state because of production over-capacity, the existence of inventories and stockpiles in various countries and because of efforts of some consumers to reduce the price of uranium to an absolute minimum, regardless of the consequences to the producers. Some sales were made at $4·00 per pound U_3O_8, less than the operating costs alone for many producers in the world.

Uranium policies originating in the United States are said[2] to have greatly increased the difficulties encountered at that time by the producing industry. Under one of the provisions of the Private Ownership of Special Nuclear Materials Act, 1964, domestic and foreign uranium customers were permitted to use the United States government-owned enrichment facilities. In order to protect the domestic U.S. uranium industry, uranium originating in other countries and enriched in the facilities was excluded from U.S. markets, which at that time represented 70% of the world market. In 1967 the United States Atomic Energy Commission (USAEC) expressed its willingness to sell to domestic and foreign customers enriched uranium based on a price of $8·00 per pound of U_3O_8 feed. This in effect set a maximum price of $8 per pound U_3O_8 on the world market.

The Canadian government made representations that the restrictions on utilisation of uranium originating in other countries was in contravention of the General Agreement on Tariffs and Trade, but without effect. On the contrary, in October 1971 the USAEC asked for comments on its plans to defer consideration of relaxation of the sale of U.S.A. enriched uranium which had originated elsewhere for use in reactors in the U.S.A., until the late 1970s. At the same time the USAEC asked for comments on its plan to sell surplus uranium stocks held by the USAEC to both domestic and foreign customers. The proposals contained detailed provisions to ensure that the domestic uranium-producing industry was not disrupted, but made no explicit provision to avoid disruption of foreign markets.

After consideration of comments on the disposal of its stockpile, the USAEC in 1972 decided to defer relaxation of restrictions on the use of enriched foreign uranium until late in the decade and to adopt the so-called 'split tails' approach for operating its diffusion plants to dispose of its surplus. The amount of uranium used in an enrichment plant depends on the amount of U235 which is left in the tails. As this is allowed to increase not only is more uranium feed required to produce a unit of U235, but the amount of separative work required is reduced. The tailings material is stored for possible recycling through the enrichment plant to recover more U235 at a later time.

The 'split-tails' approach proposed by the USAEC was to operate the diffusion plant at a tails assay in the range 0·275-0·30% U235 but to calculate the enrichment transaction on the basis of a 0·20% U235 tails assay. The additional uranium required would be supplied by the USAEC stocks and customers would be charged for more enrichment services than were in fact necessary to enrich the uranium. In fact the USAEC proposal was to exchange uranium from its stockpile for increased revenue from the sale of enrichment services. This increased revenue has been calculated as being equivalent to $10·50 to $12·00 per pound U_3O_8 for the uranium from the stockpile when the price for enrichment services was $32 per unit of separative work. Uranium was at that time selling in foreign and domestic markets in the range $5 to $6 per pound U_3O_8.

Early in 1972 the Canadian government authorised officials of two Crown corporations, Eldorado Nuclear Limited and Uranium Canada Limited to discuss with international uranium producers possible marketing arrangements for areas other than the U.S.A. They also requested private producers to attend. The discussions led to the proposal of informal marketing arrangements which excluded the United States domestic market and the domestic markets of Australia, Canada, France and South Africa.

In 1973 the largest electricity utility in the U.S.A., the Tennessee Valley Authority, invited bids from 53 producers for 86M pounds of uranium oxide for supply during the period 1979 to 1990. Only 2 producers responded and it was significant that prices quoted ranged from $12 to $16 per pound U_3O_8 plus escalation from 1973. The Arab oil embargo and the four-fold increase in the world price of oil led to the decision in several countries to expand their nuclear power programmes. This led to the development of a general attitude among uranium consumers that longer term purchasing arrangements for uranium should be negotiated and that if necessary a stockpiling policy should be adopted. At this time the USAEC proposed to relax the embargo on utilisation of foreign uranium enriched in U.S. facilities. It was apparent that there was not enough uranium to supply all the requests which were appearing in the markets.

In early 1974 Canadian producers quoted prices of $12·50 per pound U_3O_8. At about that time for various reasons curtailment of marketing activities occurred in Australia, France, South Africa and other African countries. Within the next few months future Canadian sales of 45 000 tons of U_3O_8 had been negotiated, ten times the current annual production. By the end of 1974 the world market price was $15 per pound U_3O_8, approximately twice that of a year earlier.

During 1975 the price continued to rise for several reasons. One was that the industry using the light water reactor, which was dependent on enrichment in U.S. facilities, had decided from various announcements from the USAEC that the U235 tails assay would be increased from 0·2 to 0·25% by 1977 and to 0·30 by 1981. This would require an increase of 20% in the uranium feed requirement to the enrichment plant. Another reason was a growing conviction that recycling of uranium and plutonium from spent fuel was becoming less likely, at least in the U.S.A., because of delays due to new regulations and various objectors.

In September 1975 U.S. Westinghouse announced that it would be unable to deliver 32 500 tons of U_3O_8 which it had contracted to sell within the U.S.A. and abroad. This contributed substantially to a further price increase as companies tried to cover their shortfall by entering the market and in early 1976 the price reached $40 per pound U_3O_8. By then many purchasers had accepted the new marketing concept of unpriced future deliveries. In the early 1970s it had been possible to buy uranium at fixed prices for many years into the future. Beginning in 1974 contracts were being signed based on a floor price escalating with some indices, or world market price, whichever was the higher. The price has now become relatively stable at around $40 per pound U_3O_8, responding to inflation. At this figure electricity generation from nuclear reactors remains economically viable; if it rises very much more in real terms this may not be the case.

3   Requirements for uranium and its availability

About 47% of the electricity generated from nuclear power up to the end of 1975 was produced in Western Europe. The estimated distribution of nuclear capacity in 1992 is as follows (source Nuclear Energy Agency and International Atomic Energy Agency, *Uranium resources, production and demand, including other nuclear fuel cycle data* (Paris: OECD, 1976, 78p).

	World nuclear capacity %
U.S.A.	39
Western Europe	36
Japan	8
Canada	4
Others (more than 30 countries)	13

Of the four main markets for uranium, Canada will certainly be self-sufficient and the U.S.A. will be able to satisfy most of her needs from domestic resources for some time yet. Western Europe and Japan will be the main importers of nuclear fuel. The European Community nuclear programme is about 75% of that of Western Europe and will therefore be the largest single purchaser in the world market.[3] The Euratom treaty as applied within the community requires the Euratom Supply Agency to participate in all supply contracts.

The European needs for uranium in relation to world supply and demand are of very great importance to the future market and so to countries in which uranium is produced and to the mining companies engaged in exploration and production.

Electricity utilities in Western Europe had invested in nuclear power reactors following the 1956 Suez crisis and found that such reactors often produced electricity at the least cost on their generating system. The oil crisis in 1973 again called into question the security of energy supplies and at that time the European Community rethought its energy policy and set out a series of objectives.[4] In particular it was resolved to reduce the energy dependence of the Community upon imported oil from a figure which had already exceeded 60% and was still growing, to at most 50% by 1985. One of the major ways of doing this was to undertake a massive programme of nuclear generated electricity which sought to increase its percentage contribution to the energy needs to at least 13% by 1985. In consequence electricity would satisfy more than 35% of energy requirement in 1985 in place of 25% in 1975.

The energy equivalence of the fuels available to generate electricity are indicated by the following figures derived from American sources.[5] 1 kg of U235 is equivalent to 3000 tonnes (t) of coal in energy capacity. In the U.S.A. during 1976, 4300 t of U_3O_8 equivalent was used to generate 191 billion kwh of electric power. The quantity of coal required to do this would have been 81Mt, thus 18 800 t of coal are replaced by 1 t of U_3O_8. If the nuclear power had been generated by oil or natural gas, 350 billion barrels of oil or 2 trillion cu. ft. of gas would have been needed. During 1976 the total costs of producing 1 kWh of electricity using nuclear energy averaged 1·5¢ compared with 1·8¢ for coal generation and 3·5¢ for oil. Overall, the nuclear contribution to power generation in 1976 yielded a cost saving of $1·4 billion.*

It is frequently argued that cost comparisons between nuclear reactors and coal burning for electricity generation are uncertain because of different ways in which the calculations can be carried out, but it seems to be established that at present there is little difference in the economics. Coal, however, is vulnerable to significant rises in costs due to the imposition of regulations on standards of gas emissions. Nuclear power is subject to substantial but undefined costs for permanent disposal and management of wastes, but since the cost of raw materials is only 5–10% of the total cost of nuclear power generation, it can absorb significant increases in the price of fuel and remain economically attractive. Of the 1·5¢ per kWh for nuclear generation referred to above, all steps of the nuclear fuel cycle, including provision for waste handling, accounted for 0·3¢. In the cases of coal-burning and oil-burning generation, fuel costs amounted to 1¢ of the 1·8¢ and 2·1¢ of the 3·8¢ per kWh total costs respectively.

The data for 1976 are based on the low prices for U_3O_8 which were paid up to that time. If all other charges for nuclear generation were to remain the same it would cost 1·7¢ per kWh to generate electricity with a U_3O_8 price of $40 per lb. and 2·0¢ per kWh for a U_3O_8 price of $80 per lb.

* The usage billion $= 10^9$, trillion $= 10^{12}$ has been adopted in this review.

Estimates of the total energy requirements of individual countries or areas in the future are very uncertain at present for economic reasons. Until recently the probable availability of U_3O_8 in future was equally uncertain, for political reasons, particularly in the case of Australia. The situation in 1977 was reviewed at the Second International Symposium on Uranium Supply and Demand.[6] Australia has now decided to mine uranium, under strict controls and, like Canada, to supply U_3O_8 if the required safeguards and restrictions are agreed to by the customer.

The facts concerning known reserves of uranium ore at prices for U_3O_8 of $30 and $50 per lb. are shown in Table 1. It should be noted that a material containing

Table 1. Uranium resources (000 short tons U_3O_8)

	Reasonably assured		Estimated additional	
	$30 per lb. U_3O_8	$50 per lb. U_3O_8	$30 per lb. U_3O_8	$50 per lb. U_3O_8
United States of America	690	890	1 120	1 450
South and S.W. Africa	400	450	44	94
Australia	380	380	60	60
Canada	215	240	510	850
Niger	210	210	69	69
France	48	67	31	57
India	39	39	31	31
Algeria	36	36	65	65
Gabon	26	26	7	13
Brazil	24	24	11	11
Argentina	23	54	0	0
Central African Republic	10	10	10	10
Japan	10	10	0	0
Spain	9	9	11	11
Portugal	9	11	1	1
Yugoslavia	6	8	7	27
Mexico	6	6	3	3
Turkey	5	5	0	0
Germany	2	3	4	5
Zaire	2	2	2	2
Italy	2	2	1	1
Austria	2	2	0	0
Sweden	1	390	4	4
Greenland	0	8	0	11
Somalia	0	8	0	4
Finland	0	4	0	0
Korea	0	4	0	0
United Kingdom	0	0	0	10
Chile	0	0	0	7
Madagascar	0	0	0	3

Sources: U.S. Department of Energy, January 1978, OECD Nuclear Energy Agency and International Atomic Energy Agency, December 1977.

uranium does not become a reserve of ore until it is economic to recover uranium from it. Known production capacity for U_3O_8 planned for 1977, 1980 and 1985 is shown in Table 2.

Table 2. Production capacity for U_3O_8 (short tons)

	1977	1980	1985
United States of America	16 100	26 900	39 500
South Africa	8 710	15 200	16 200
Canada	7 930	10 400	16 300
France	2 860	3 700	4 820
Niger	2 090	5 330	11 700
Gabon	1 040	1 560	1 560
Australia	520	650	15 300
Spain	250	880	1 650
India	260	260	260
Argentina	170	470	780
Germany, Fed. Rep.	130	130	260
Portugal	110	120	350
Japan	39	39	39
Brazil	0	500	500
Mexico	0	220	720
Yugoslavia	0	160	230
Italy	0	160	160
Turkey	0	130	130
Philippines	0	49	0
Central African Republic	0	0	1 300
TOTALS	40 209	66 858	111 759

Sources: U.S. Department of Energy, January 1978, OECD Nuclear Energy Agency and International Atomic Energy Agency, December 1977.

4   The mode of occurrence of uranium

According to the classification of Goldschmidt, which is generally accepted by geochemists, uranium is a lithophile element which combines in nature with oxygen. It can occur as U(IV) or as U(VI) and as the uranyl cation UO_2^{2+}. The common rock-forming minerals do not easily accommodate the element because it cannot substitute for any other cation of its own size, for example calcium or sodium, as the balancing of excess electronic charges is practically impossible. Consequently, when the minerals were being formed from magma, the uranium present was either concentrated in the late magmatic fluids or expelled from the magma chamber.

When separating from magma, the tendency is for uranium to form its own minerals such as uraninite (an oxide) or to somehow become associated with accessory minerals. In granites, which contain low concentrations of uranium, this may be dispersed as ionic or molecular disseminations. These molecules are usually found in crystal structures, adsorbed on crystal surfaces, in lattice defects, or in fluid inclusions. Therefore the tendency will be for uranium to be progressively concentrated within a differentiated suite of ingeous rocks. The question of how uranium precipitates from a magma is not fully resolved, but most uranium-bearing granites seem to be of similar type and composition and were formed under certain specific, stringent conditions. Acid igneous rocks such as granite have distinctly higher concentrations of uranium (4ppm) than basic igneous rocks (1ppm).

Estimates of the average abundance of uranium in the earth's crust vary between about 2 and 4 ppm, similar to values for beryllium, arsenic, molybdenum and tungsten, but higher than those for silver, cadmium, bismuth and mercury. The average abundance of uranium in the geochemical cycle is shown in Table 3. Recovery of all elements having this order of abundance depends on their having been concentrated in some way either during deposition from a magma or by a secondary process.

Table 3. Average abundance of uranium in the geochemical cycle.

Host Environment	ppm U
Mean value in lithosphere	2·6
Ultrabasic rocks	0·002
Basalts	0·75
Syenites	3·0
Diorites	1·8
Acid rocks (high Ca)	3·0
(low Ca)	3·0
Slates and clays	3·2
Sandstones	0.45
Carbonates	2·2
Sea water*	0.003

Sources: Rösler, H. J. and Lange, H., Geochemical tables. Elsevier, Amsterdam, 1972.
* Krauskopf, K. B., Introduction to geochemistry. McGraw-Hill, New York, 1967.

In the case of uranium, during weathering of the rocks containing it, the element was leached out under oxidising conditions and moved with the ground water. This could contain anions such as sulphate and carbonate which stabilise the uranium in solution by the formation of complex species, and others such as fluoride, phosphate, carbonate, chloride and silica (in various forms) which could become associated with uranium in forming insoluble solids, uranium minerals. It seems likely that preconcentration was an essential step in the formation of many uranium ores and this probably took place by adsorption of ions on materials such as clays and oxyhydroxides of ferric iron, manganese and titanium.

Precipitation of uranium from solutions to form sedimentary ore deposits can take place by ion exhange, neutralisation, chemical replacement, or by reduction. The latter is the process responsible for the formation of the minerals containing U(IV) which are particularly important in ores. The reducing agent may be carbonaceous matter or hydrogen sulphide among others. Anaerobic bacteria which reduce sulphur produce hydrogen sulphide by feeding on organic matter and sulphates. Sulphides such as those of iron, pyrite, marcasite and pyrrhotite, which are associated with uranium minerals in many of the major sandstone-type deposits in the U.S.A. and elsewhere show evidence of having been formed by the action of bacteriogenic hydrogen sulphide.

Thus the main factor which controlled the distribution of deposits of uranium appears to have been the early evolution of the earth's crust. The oldest known uranium minerals concentrated in economic quantities, in Precambrian sediments, are more than 3000M years old. Well over 90 % of the known uranium deposits which can be worked at low cost occur either in Precambrian rocks or in Phanerozoic rocks immediately overlying the basement.

By measuring the ratios of uranium, thorium and lead present using a mass spectrometer it is possible to estimate the date at which uranium mineralization occurred with considerable precision. Information obtained in this way shows that uranium deposition generally took place over a long period of time, mostly between 2000M and 600M years ago.

The main exceptions are the deposits of the Colorado-Wyoming province of the U.S.A. which are essentially late Cretaceous to early Tertiary in age, although they appear to have been formed between 210M and 10M years ago and some may even be forming at present. These deposits are, however, close to the Precambrian basement which is a southwestern extension of the Canadian shield.

Major uranium provinces can be considered for convenience to be those with more than 100 000 tonnes of uranium reserves. The four known at present are Elliot Lake-Blind River, Canada; Colorado-Wyoming, U.S.A.; Witwatersrand, South Africa; and Northern Territory, Australia. Most of the reserves occur in sandstones, vein-type deposits are second in importance, conglomerates are third and other deposits (excluding shales) fourth. The shales of southern Sweden constitute a special case; although reserves are large annual production from them is not likely to exceed 1000 tonnes of uranium.

4.1   Occurrence in the uranium provinces

4.1.1   Canada.
By far the largest proportion of Canadian uranium occurs in quartz-pebble conglomerates of Precambrian age in the Elliot Lake-Blind River district of Ontario. The main ore minerals are uraninite (ideally UO_2) and brannerite, AB_2O_6, (where A is mainly U but also Ca, Fe, Th, Y; B is mainly Ti but also

Fe) which occur as grains 100 μm or so in diameter. In the Agnew Lake area appreciable thorium also occurs in the minerals uranothorite (U, Th)O_2.SiO_2 and monazite MPO_4 (where M is Ce, Nd, Pr, La and includes some U) and this assemblage makes the ore more difficult to treat.

With further deposits being discovered frequently, Northern Saskatchewan is emerging as a major uranium-bearing area. The ore at Cluff Lake is of such a high grade, containing 4·25% U in one zone, that some tailings will have to be impounded in concrete casks owing to their high radioactive content of decay products, largely radium.

4.1.2   U.S.A. About 90% of the known ore reserves in the U.S.A. are in the Colorado-Wyoming province. The uranium mineralisation occurs mainly in sandstones and mudstones ranging in age from Triassic to Eocene. Unlike the situation in Canada, where there are a few large deposits, in the Colorado-Wyoming province there are more than 1000 deposits with ore-grade material, although probably 90% of the reserves are contained in 10% of the deposits.

More than 30% of the ore is expected to be mined by open-cut methods and this distribution of uranium gives the U.S.A. a flexibility in rate of production which has not been available elsewhere. The primary uranium minerals present are uraninite and coffinite, $U(SiO_4)_{1-x}(OH)_{4x}$ which are relatively easy to treat.

These facts together make it possible to recover the uranium present in some of these deposits by *in-situ* leaching, a field of multi-discipline technology which is being very actively developed at present. This is dealt with below.

4.1.3   South Africa. The uranium deposits of the Witwatersrand Basin are the most important in South Africa, although a large quantity of uranium occurs in the deposits of Rössing near the edge of the Khan River basin to the north east of Swakopmund, South West Africa (Namibia).

The deposits of the Witwatersrand are among the oldest known, 2000M years or more in age, and in many ways are similar to those which occur in the Elliot Lake-Blind River province. However, they contain only about 300 ppm U_3O_8 as opposed to 1200 ppm mean grade in the Canadian deposits and until recently could generally only be worked economically as a by-product of gold recovery. The main ore mineral is uraninite which occurs in grains averaging only about 10μm in diameter. More finely divided uranium phases also occur in association with phyllosilicates and as a result there is an appreciable loss of uranium into the slimes fraction during treatment.

4.1.4   Australia: Exploration practically ceased in Australia between 1955 and 1965 but was then revived by companies encouraged by the prospective improvement in the uranium market. Political moves to take over the uranium industry, between 1972 and 1975, brought exploration to a standstill again and prevented development of the newly discovered deposits. Exploration has continued again since 1975 and the overall result has been a spectacular increase in the known reserves from 71 000 t U in 1973 to 243 000 t in 1975 and a further rise since, Table 2. Most of the additions are in the uranium province of Northern Territory which was discovered in 1949. The South Alligator River area contains deposits which were worked in the 1950s and includes El Sherana, which contained massive pitchblende concentrations such as had been known previously at the famous Shinkolobwe mine in the then Belgian Congo. The recent and even more spectacular discoveries in the Alligator Rivers area also contain pitchblende which, being a colloform, fine-grained variety of uraninite, is amenable to treatment. Some deposits contain refractory minerals such as brannerite and davidite, $A_6B_{15}(O,-OH)_{36}$, (where A is Fe^{2+}, rare earths, U, Ca, Zr, Th; B is Fe^{3+}, Ti, V, Cr) which will make processing more difficult. The ore is in general of higher grade than that found elsewhere.

Most of the uranium deposits in Australia including those in the Alligator Rivers region, can be classified as vein-type deposits. However there are sedimentary deposits in South Australia which somewhat resemble those in the Colorado-Wyoming province. It may be practicable to recover uranium from some of these by in situ leaching. There are other, unusual, deposits in Australia, including the Maureen where uranium-fluorine-molybdenum mineralisation is found in a basal sandstone.

Australia has large deposits of uranium in South and in Western Australia and so is a country of major importance as having a capability of producing very large tonnages of low-cost uranium. This fact could be a controlling factor of the world market in the 1980s. Indeed, the uranium ore reserves in the Northern Territory province could be five to ten times as much as has been discovered so far.

A general discussion of uranium deposits and mineralogy has been given by Merritt,[7] and more detailed information on some areas is contained in several papers in reference 3.

5   Uranium extraction

5.1   The chemistry used

The inorganic chemistry of uranium is very complicated but the aspects made use of in recovering the element from its ores are straightforward. A recent review[8] contains a compilation of thermodynamic data for aqueous uranium-containing species, oxides and some minerals, and deals with equilibria in solutions and between solutions and minerals.

Although there is a U(III) species which can occur in aqueous solutions the important valency states are U(IV) and U(VI). The former can occur in aqueous solutions as U^{4+} ions whereas the latter exists as the uranyl ion UO_2^{2+}.

The solid oxides of uranium exist over a range of stoichiometry and form double, or more complicated, oxides with several other metals. Many such oxides have precipitated from ground water, and more than 170 uranium minerals have been described. Very few can be classed as important ore minerals however. Several of these have been mentioned above as providing the main uranium mineralisation in the uranium provinces, and they mainly contain U(IV). Among the large number of uranyl, or U(VI) minerals, the vanadates carnotite, $K_2(UO_2)_2(VO_4)_2.3H_2O$, and tyuyamunite, $Ca(UO_2)_2(VO_4)_2.5\text{–}8H_2O$, the phosphates of the autunite group, $Ca(UO_2)_2(PO_4)_2.10\text{–}12H_2O$, and the silicate uranophane, $Ca(UO_2)_2(SiO_3)_2(OH)_2.5H_2O$ are the only ones found in large enough quantity to be classified as ore minerals.

The regions of pH and oxidation potential in which simple uranium oxides, ions in solution and insoluble uranates occur are shown in Fig. 1. Values of E are on the hydrogen scale and larger positive values represent more oxidising conditions.

Leaching of uranium minerals involves selecting a region of pH and oxidation potential where the solids are unstable and the ions containing uranium are stable in solutions. It is clear from Fig. 1 that acidic solutions will dissolve uranium oxides and under oxidising conditions will form the soluble uranyl salt. In practice use of an oxidising agent is necessary to cause minerals containing U(IV) to react and dissolve. The uranyl minerals do not require an oxidising agent but the pH of the leaching solution must be outside the region over which they are sparingly soluble.

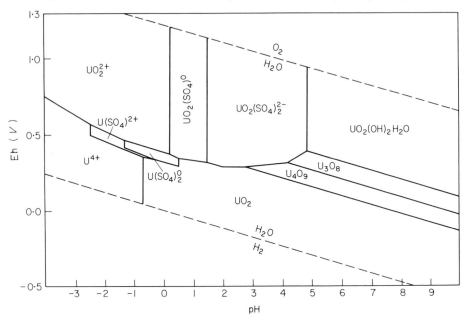

Fig. 1. E-pH diagram for $U-S-H_2O$ at 25°C. Activities of U-containing ions, 10^{-2}; of S-containing ions, 10^{-1}. Courtesy C. J. Hallett, Imperial College, London.

It can also be seen from Fig. 1 that if the pH of a uranium-containing liquid is raised an insoluble uranate is precipitated and such solids are known as 'yellow cake', the product containing uranium which is sold from the mine site. Two process routes are available, therefore, for leaching uranium from its ores, acid leaching, usually under oxidising conditions, and alkali leaching over a restricted range of pH, usually achieved using sodium carbonate-bicarbonate or, in the case of *in situ* leaching, the ammonium salts.

Acid leaching in particular dissolves many elements besides uranium which are present in the ore. Effective separation of the uranium from most of the impurities can be achieved by making use of the fact that the uranyl ion forms stable complexes with oxyanions such as sulphate, carbonate and nitrate. In the case of sulphate the equilibria are

$$UO_2^{2+} + SO_4^{2-} \rightleftharpoons UO_2SO_4$$

$$UO_2SO_4 + SO_4^{2-} \rightleftharpoons UO_2(SO_4)_2^{2-}$$

$$UO_2(SO_4)_2^{2-} + SO_4^{2-} \rightleftharpoons UO_2(SO_4)_3^{4-}$$

Thus by controlling the concentration of sulphate, carbonate or nitrate in the uranium-containing solution the uranium can be made to exist as a cation, uncharged species, or an anion. Removal from solution as an anion by an ion exchange resin or by liquid liquid extraction (solvent extraction) using an amine in a diluent, generally gives excellent separation of uranium from most impurities.

In the presence of significant concentrations of sulphate or carbonate ions the complex uranium-oxyanion species are much more thermodynamically stable than the simple UO_2^{2+} ion so that the regions of stability of the oxides are much smaller than in the absence of sulphate or carbonate. Thus the presence of these ions during leaching increases the driving force of the leaching reaction, independently of the effect of pH.

After the uranium has been eluted from the resin or stripped from the solvent it is usually precipitated as yellow cake from the resulting aqueous solution by adding alkali, very often ammonia although magnesia or sodium hydroxide may be used. In cases where this does not give a product of acceptable quality or, as is becoming increasingly likely, when the use of ammonia causes difficulties over discharge of effluent, hydrogen peroxide can be used to precipitate the uranium.

In slightly acidic solutions uranyl salts and hydrogen peroxide undergo a reaction analogous to hydrolysis with precipitation of hydrated UO_4

$$UO_2^{2+} + H_2O_2 = UO_4 + 2H^+$$

The method is highly selective for uranium and precipitation of uranium is practically quantitative in the pH range 2·5–4·0 with some excess hydrogen peroxide present.

5.2   Extraction processes

A comprehensive account of the processes used in the extraction of uranium from its ores and of unit operations involved, as well as the flowsheet of each active, inactive and past uranium plant in the U.S.A., was published in 1971.[7] This includes essentially the whole of the technology which was used in the field up to that time.

All processes for recovering uranium from ores involve leaching and this step is of fundamental importance to the overall flowsheet since it involves a choice between using acid or alkaline conditions. Sulphuric acid leaching is used when possible because it is more effective with most ores and does not require the ore to be ground so finely as does carbonate leaching. This is because the acid attacks other minerals as well as those containing uranium and so gains access to uranium inside the individual particles. Several new developments in leaching methods make use of this fact. Carbonate leaching, on the other hand, is more selective than acid leaching and is particularly useful for treating ores which would consume large quantities of acid, such as those containing much lime. Simplified flowsheets for typical acid and carbonate processes are shown in Figs 2 and 3.

If the uranium minerals contain U(IV) an oxidant is added during leaching. This may also be desirable if the ore contains no U(IV) but does contain reducing agents, including metallic iron originating from mining and comminution stages, which cause difficulties. Reagent concentrations, pulp density, leaching time, oxidation potential and temperature are the variables which are controlled.

The two important techniques which are used for concentration and purification of the uranium are resin ion exchange and solvent extraction. Strong base anion exchange resins are used which preferentially adsorb anionic uranium complexes present in the leach solution, and exclude metal cations. These resins are used to treat both acid and alkaline leach liquors. Solvent extraction is used to treat clarified acid liquors and both anionic and cationic extractants are being used. Solvent extraction has been used also after resin ion exchange, as a second stage of purification, to obtain very high grade products.

Chemical precipitation is used to recover the uranium from solution after purification. In alkaline circuits sodium hydroxide is added to raise the pH to nearly 12, when a sodium uranate is precipitated. In acid process plants the eluate from ion exchange resin or strip liquor from solvent extraction is treated with an alkali to precipitate a uranate. Usually ammonia or magnesia are used although lime may be employed for partial neutralisation in the first step of a two-stage process to remove iron and other impurities still present in the solution, which can be precipitated at a lower pH than uranium. A high grade product can also be made by precipitating uranium peroxide using hydrogen peroxide.

Final stages in the production of uranium materials from ores include dewatering, drying and packaging. Calcining, or drying at high temperatures is effective in removing ammonia, sulphate and other volatile constituents, but temperature control is very important to avoid making the product refractory to subsequent refining processes.

Recent developments in processes for extracting uranium from ores are concerned with changes in flowsheets resulting from changes in individual stages of the overall process. It is, therefore, convenient to deal with these stages individually. In spite of the high capital and operating costs of crushing and grinding the ore these steps are not considered here. It should be noted however that in order to leach uranium from a particle of ore it is necessary only to expose the uranium

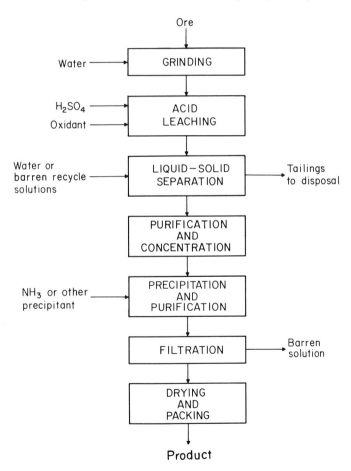

Fig. 2. Outline generalized sulphuric acid leach process for uranium.

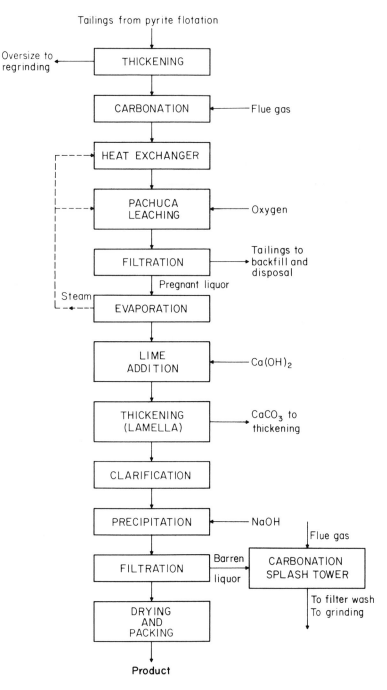

Fig. 3. Outline carbonate leach process (Eldorado) for uranium.

mineral, not to liberate it. If on the other hand a concentrate is to be produced by a physical process, such as froth flotation or a method depending on specific gravity, then liberation of the minerals to be concentrated will be essential. In the case of uranium recovery from gold ores the degree of comminution is controlled by the gold circuit.

5.2.1   *Carbonate leaching.* In the past the carbonate process was selected when the sulphuric acid consumption of the ore exceeded 100 kg/tonne, which is equivalent to about 7–9% calcite in the ore. Other factors must now be considered in addition, and carbonate leaching has advantages over acid leaching on environmental, health and safety, and water usage grounds. In spite of this, the vast majority of the uranium mills in the world use the acid leach process. In Canada only Eldorado's Beaverlodge mill uses the carbonate process, and this accounted for less than 10% of Canadian output in 1978.

The carbonate process as shown in Fig. 3 requires the recovery and recycle of reagents. Sulphides present in the ore are removed by froth flotation before leaching, and uranium contained in the concentrate is recovered by acid leaching.

The reactions involved in the alkali leaching process can be represented as

$$UO_2 + \tfrac{1}{2}O_2 = UO_3$$

$$UO_3 + 3Na_2CO_3 + H_2O = Na_4UO_2(CO_3)_3 + 2NaOH$$

In order to prevent the pH rising to a value at which reprecipitation of uranium will occur, sodium bicarbonate is used to buffer the hydroxide formed.

The precipitation reaction is

$$2Na_4UO_2(CO_3)_3 + 6NaOH = Na_2U_2O_7 + 6Na_2CO_3 + 3H_2O$$

and after removal of the sodium diuranate the excess sodium hydroxide in the barren solution is converted to carbonate and bicarbonate by contacting the solution with boiler flue gas.

The rate of leaching of oxides containing U(IV) in alkaline solutions is controlled by the oxidation of this to U(VI) and in the case of reaction with oxygen an electrochemical mechanism has been proposed.[9] At the Beaverlodge plant, which has been in operation since 1953, 99·5% oxygen is used as the oxidant, produced from an oxygen plant of 9 tonnes per day capacity.

The nominal capacity of the uranium mill is 1650 tonnes per day and the average grade of the ore reserves is 0·21% U_3O_8. A pyrite concentrate is taken by froth flotation and the tailings are thickened to 55% solids and pumped to the carbonation system where two high-capacity Outokumpu flotation agitators are being adapted for dispersion of the boiler house flue gases. The carbonated pulp passes through a series of three concentric heat exchangers to raise the temperature to 90°C. The leaching circuit consists of twenty-four 9000 cu. ft. pachuca tanks in four banks of six. Mechanical agitators driven by 15 hp mixers pull upwards in 16 inch and 24 inch diameter tubes. The oxygen used to oxidise the uranium and residual sulphide minerals is sparged deep into each vessel. Retention time is four days and uranium recovery is above 92%.

The introduction of the neutralisation step using lime, which is shown in Fig. 3, lowered the content of Ra226 in the yellow cake produced to about 10% of its previous level. The Ra226 in the tailings is mainly associated with the −0·076 mm fraction. The tailings are pumped through 8 inch plastic pipes to a Dorrlone plant where the + 0·076 mm solids, comprising about 50% of the total, are separated and returned underground for backfilling. The tailings slimes are sent to a settling pond in the tailings management area. The overflow from this pond is treated with barium chloride and flocculants to remove radium, before the water is discharged to Beaverlodge Lake.

5.2.2   Sulphuric acid leaching. In most uranium ores the element will be present in several, usually many, different minerals. Some of these will dissolve in sulphuric acid solutions under mild conditions, others require more severe conditions. Thus it may be easy to recover 90–95% of the uranium present, but difficult or impossible to obtain the remaining few percent economically. Considerable efforts are being made to achieve increased recoveries by a number of changes in leaching methods. Some of the most difficult uranium minerals to leach are those of the multiple oxide type, most commonly brannerite and davidite. These usually contain U(IV) as well as U(VI), together with such elements as titanium, iron, vanadium, thorium and rare earths. To recover uranium from them it may be necessary to leach for 60 hours at 77% solids with 50 g/l of free sulphuric acid, maintained throughout the period, at 70°C, as was done at one mine in the Elliot Lake district of Canada, Sodium chlorate was used as the oxidant. Minerals containing uranium together with niobium or zirconium are even more difficult to treat.

In a normal acid leaching plant in which ore is agitated with acid and an oxidising agent in pachuca tanks the principal requirement is to maintain a free acid concentration sufficient to attack the uranium minerals without dissolving an excessive quantity of the gangue minerals present. At the end of the period of leaching sufficient acid must remain to prevent reprecipitation of the uranium. Temperature and the concentrations of the varous constituents of the solutions will determine the actual pH at which precipitation of a compound will occur, but if U(IV) remains in solution because of the presence of reducing minerals when all of the oxidant has been consumed, a phosphate and an arsenate will precipitate at a pH below 1·2. Uranyl arsenate precipitates in the range pH 1·3–1·7 and the phosphate between pH 1·9 and 2·5. The possibility of causing reprecipitation of uranium when washing the tailings must be borne in mind also.

Maintenance of proper oxidising conditions during leaching is next in importance to acid concentration in obtaining a high recovery of uranium. The rate controlling process in acid leaching of at least some minerals containing U(IV) has been convincingly shown to be electrochemical in nature and to depend on the nature of the electron transfer processes occuring in both the anodic and cathodic reactions.[10] The oxidation reaction occurs at the surface of the uranium mineral and there are very few materials capable of performing this oxidation. The only substances used for this purpose at present are ferric salts, oxygen and hydrogen peroxide (for *in situ* leaching).

The two oxidising agents most commonly used in uranium leaching are manganese dioxide (as pyrolusite) and sodium chlorate but neither of these oxidises the U(IV). They both oxidise ferrous to ferric ions. Thus when either is used the oxidising conditions within the leaching pulp can be monitored by using the redox potential as measured by an inert electrode. The value of the potential is controlled by the Fe^{3+}/Fe^{2+} ratio in the solution. The concentration of Fe^{3+} necessary to achieve an adequate rate of leaching is usually more than 0·5 g/l. and may be as high as as 5–6 g/l. Most uranium ores, although not all, contain sufficient iron minerals to provide sufficiently high concentrations by reaction with the acid leach solution with no oxidant present. This is added after reducing gases such as hydrogen and H_2S, caused by reaction with metallic iron and iron sulphides, have ceased being evolved. The average amount of manganese dioxide used is about 5 kg/tonne ore, often as 10 kg of 50% pyrolusite/tonne. Up to about 1·5 kg of sodium chlorate/tonne of ore is generally used.

In the case of *in situ* leaching of uranium, ferric salts cannot be used as oxidant under alkaline conditions and may be unacceptable for environmental or practical reasons in acid leaching conditions. Hydrogen peroxide or oxygen injected into the strata under pressure are used, therefore. Peroxymonosulphuric acid (Caro's acid) does not catalytically decompose in the presence of metal ions in solution as rapidly as does hydrogen peroxide and its use as an oxidant in uranium leaching is being examined.[11]

Time and temperature of leaching are interdependent but since the cost of heating is high and the leaching tanks used are simple it is common to use ambient temperatures and longer retention times. In the U.S.A. leaching times generally are between 4 and 24 hours at temperatures between ambient and 80°C. In Canada and South Africa leaching times may be up to 48 hours. The pulp density used is usually the maximum possible which gives good recovery of uranium.

Solid–liquid separation after leaching relies heavily on the use of flocculants to accelerate the rates of settling and filtration of the ores which generally contain clays and other slimy constituents. The development and use of such materials as the guar gums and polyacrylamide polymers had very great effects on uranium extraction technology and made it possible to filter and thicken some ores which otherwise would have been very difficult to process. In South Africa and Canada filters in multiple stages are preferred, sometimes combined with thickeners, while in the U.S.A. countercurrent decantation in thickeners is generally used. Hydrocyclones are often used to take cuts for particular purposes.

The long leaching times and high free sulphuric acid concentrations referred to above for the treatment of an ore from the Elliot Lake region lead to a sulphuric acid consumption of 30–40 kg/tonne of ore. This can be lowered by about 10 kg/tonne if acid recycle or 2-stage leaching is used, the pulp being 60–70 % solids, which is very thick. In the UKAEA strong acid process for treating refractory ores[12] a similar quantity of sulphuric acid, at a concentration which may be about 3M is mixed with the ore to dampen it to about 90 % solids and the mass is cured at the desired temperature. This proposed method of decomposing the structure of the uranium minerals is analogous to the pugging process which was used in the U.S.A. and is at present used to treat at least one ore in Niger. In the latter case, after curing, or baking, the mass is digested with water and oxidised using sodium chlorate. This ensures that both uranium and the molybdenum present are fully oxidised so that solvent extraction methods can be used to recover both.

In South Africa, when the process for recovery of the uranium present in some gold ores was developed, the tailings from the cyanidation plant were washed, filtered, re-pulped and then leached with sulphuric acid. One consequence of this was that the small amount of cyanide present in the leach solution reacted with the little cobalt which was leached from the ore to form the cobalticyanide ion $Co(CN)_6^{3-}$. At that time strong base anion exchange resins were used to extract the uranium from the leach liquor and these also adsorbed the cobalticyanide ions. It is impossible to elute these effectively from the resin which gradually lost its ion exchange capacity and eventually was completely poisoned. This was one major reason for installing solvent extraction for uranium in place of ion exchange.

At the Hartebeestfontein mine the so-called 'reverse leach' circuit was introduced in which the milled ore was first acid leached to recover the uranium and subsequently leached with cyanide to extract the gold. This apparently simple change had very marked effects on the possibilities available in the overall process. Some of these are as follows:

(i)   There is a small but economically significant increase in gold extraction. This is attributed to the acid attack on secondary sulphides and silicate minerals exposing gold which was contained within particles of ore, and to the effect of the acid in cleaning up tarnish on gold grains and so making them more amenable to cyanidation.

(ii)   Avoidance of formation of cobalticyanide. The original resin was still in use after about twenty years of continous use, although its capacity was slowly falling.

(iii) Avoidance of the formation of ions such as thiocyanate which interfere with the autoxidation reaction in which the ferrous iron in the leach liquor from which the uranium has been removed is oxidised to ferric by reaction with air and sulphur dioxide.

$$2FeSO_4 + 2SO_2 + H_2O + 1{\cdot}5O_2 = Fe_2(SO_4)_3 + H_2SO_4$$

(iv) A pyrite concentrate can be obtained by froth flotation before leaching. Xanthate inhibits cyanidation of gold but acid treatment for uranium recovery destroys it before the ore passes to the gold circuit.

The pyrite concentrate may represent only 2% or so of the mass of the ore but contain about 80% of the gold and 35% of the uranium present in the ore. It can thus be ground very finely and given special treatment so that 98·5% recovery of gold is achieved compared with 95% recovery from the flotation tailings and by conventional treatment of the whole ore. The pyrite, after cyanidation, is used to manufacture sulphuric acid and the calcine is used to replenish the iron content of the leach liquor, any uranium in the calcine used for this purpose being recovered. The gold in this calcine is recovered when the solids pass to the cyanidation circuit. Much of the uranium and gold which is not recovered in normal treatment plants is contained in the mineral thucholite which is a complex of uraninite with hydrocarbons, uraniferous auriferous coal particles. This material reports in the concentrate during froth flotation and recovery of gold from it is considerably increased by the fine grinding and special treatment which it is subjected to.

5.2.3   Pressure leaching. Uranium ores which contain sufficient sulphide minerals can be leached without the addition of any reagent except oxygen by using water at high temperature and pressure. This process was investigated in 1954[13] and it has subsequently been shown that it can be used to treat many Canadian, South African and Australian ores.

The chemistry of the process can be represented by considering pyrite and uraninite although a real system will be much more complex:

$$2FeS_2 + 7{\cdot}5O_2 + H_2O = Fe_2(SO_4)_3 + H_2SO_4$$

$$UO_2 + Fe_2(SO_4)_3 = UO_2SO_4 + 2FeSO_4$$

$$4FeSO_4 + 2H_2SO_4 + O_2 = 2Fe_2(SO_4)_3 + 2H_2O$$

The reactions proceed at a sufficiently rapid rate at temperatures of about 160–200°C and an oxygen partial pressure of 20–50 psi. Under these conditions in a full scale continous commercial plant treating Witwatersrand ores a retention time of

1–4 hours would generally be necessary if 2–5 kg/tonne of sulphuric acid were added to the pulp to initiate the reations, or rather longer if no acid were added.

At temperatures around 200°C hydrolysis of many metal salts is much more pronounced than at lower temperatures, particularly in the case of trivalent metals, so that additional acid will be produced by this means. It seems likely therefore, that dissolved oxygen will play an increasing role in oxidising U(IV) directly as the solubility of ferric iron decreases with rising temperature.

The high temperature used in pressure leaching leads to efficient leaching and high recovery of uranium from refractory minerals such as brannerite and others of the mixed oxide type, but thucholite is not effectively attacked. The gold content of the ore is also cleaned and made accessible by pressure leaching to a greater extent than by conventional acid leaching, and significantly more gold is recovered by cyanidation after pressure leaching than in normal reverse leaching.

The reason why pressure leaching has not been used so far in uranium extraction is, of course, that there are serous problems involved in pumping slurries to these high (by hydrometallurgical standards) pressures, operating large autoclaves under these conditions and releasing the slurries back to ambient conditions, and in recovering the heat from the pulps. In addition, acidic ferric sulphate solutions are very corrosive, and strongly agitated suspensions of particles of hard rock are very abrasive. However, a large pressure leach pilot plant was built in South Africa during 1976, capable of treating a 60% solids pulp at the rate of 20 tons of solids per hour, with a residence time of 2 hours. The autoclave used is a horizontal unit divided into four compartments by baffles, and each compartment has a mechanical agitator. The length of the autoclave is about 13 metres and its diameter 3 metres. It is of mild steel construction lined with an asbestos membrane followed by a lead lining and then acid-proof brick. Test work has also been undertaken using a pipeline reactor of the type which is in use for the digestion of bauxite in the Bayer process for alumina.

The introduction of continuous pressure leaching of acidic slurries at high temperatures would be a major advance in the technology of hydrometallurgy, and could have far reaching consequences in the extraction of a number of metals.

5.2.4   Solvent extraction. The most commonly used solvent extraction process for the recovery of uranium from leach liquors is the Amex process. The extractant is usually trioctylamine which is typically employed as a 5% solution in kerosine, or some other mainly aliphatic hydrocarbon, with 2% isodecanol present to increase the solubility of the amine salts in the diluent. The chemistry involved can be written as follows, where species with a bar above the formula are present in the organic phase

$$2\overline{R_3N} + H_2SO_4 \rightleftharpoons \overline{(R_3NH)_2SO_4}$$

$$\overline{(R_3NH)_2SO_4} + H_2SO_4 \rightleftharpoons 2\overline{(R_3NH)HSO_4}$$

$$2\overline{(R_3NH)_2SO_4} + UO_2(SO_4)_3^- \rightleftharpoons \overline{(R_3NH)_4UO_2(SO_4)_3} + 2SO_4^{2-}$$

Each reaction is reversible with an equilibrium constant.

Usually several stages of extraction are used in a countercurrent cascade, with mixer-settlers. The aqueous raffinate may be returned to the leaching circuit if traces of entrained solvent are removed to a sufficient extent by froth flotation or some other method. Uranium is stripped from the loaded organic phase using one of several possible aqueous solutions, depending on what impurities are in the organic phase and what method is to be used in precipitating the uranium from the strip liquor. The solvent circuit is often used to increase the concentration of uranium in the strip liquor to a higher value than in the leach liquor. A generalised flowsheet for the Amex process is shown in Fig. 4. The stripping solution may be (i) a sulphate solution, e.g. $1.5M$ $(NH_4)_2SO_4$; (ii) Na_2CO_3 solution, e.g. $0.75M$; (iii) MgO slurry, e.g. 20 g MgO/1; (iv) nitrate solution, e.g. $0.9M$ NH_4NO_3–$0.1M$ HNO_3; (v) chloride solution, e.g. $1.0M$ NaCl–$0.05M$ H_2SO_4.

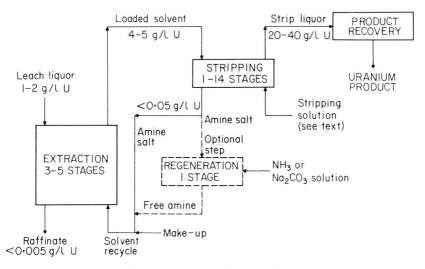

Fig. 4. Amex process for uranium extraction, generalised flowsheet.

The following description would be realistic for a plant using the Amex process, with 60 000 tonnes of ore passing through the uranium plant per month. After leaching with sulphuric acid the pulp is filtered in 2 stages, when the liquid contains about 200 ppm suspended solids. It is held in a settling tank, also acting as a surge tank, and when liquid is taken to the solvent extraction plant it contains 150 ppm solids. If an aqueous phase containing more than about 10 ppm solids is used in mixer settlers the solid particles form 'crud' (Chalk River Unidentified Deposit) which is an intimate mixture of solid, aqueous phase and organic phase, which collects at the interface between the two liquids in the settlers and seriously affects the separation characteristics. The liquid is, therefore, clarified by filtration so that it contains 7–10 ppm solids. A sand filter may be used for this.

The extractant is 5% Alamine 336 + 2% isodecanol in kerosene and the flow rate is 1800–2000 litres/minute of pregnant solution with an organic: aqueous phase ratio 1·15:1. There are 4 stages of extraction which is operated organic continuous. (This minimises entrainment of droplets of organic phase in the aqueous raffinate which is recycled or discarded.) The final organic loading is 5 g/l and the final raffinate concentration 0·002 g/l U. The total loss of organic phase by entrainment, solubility, evaporation etc. is 100 ppm of aqueous phase treated. The solvent inventory is 160 000 litres and the amine loss is 5–15 ppm of aqueous phase treated.

The loaded organic phase is scrubbed with water to remove as much sulphate and silica as possible in 5 countercurrent stages. Stripping is in 12 stages. Stages 11 and 12 contain conductivity probes to ensure that they run aqueous continuous, and the pH is adjusted using ammonia. If these stages flipped to organic continuous there would be a continous demand for ammonia. The pH in stage 11 is maintained at 5 and in stage 12 at 5·5. The strip liquor contains 3–4 g/l U_3O_8 and gaseous ammonia is used to precipitate ammonium diuranate at pH 7·1.

The amine is converted from the sulphate form to free base by regeneration with sodium carbonate before return to the extraction step.

Other extractants are sometimes used to extract uranium in cationic form, especially di(2-ethylhexyl) phosphoric acid which may also be mixed with tributyl phosphate to exploit a synergistic effect. The selectivity for uranium is not as good as that of amines, but they will efficiently extract uranium from dilute acidic sulphate, nitrate and chloride solutions in which the uranium is largely present in cationic species. When ferric iron is present the phosphates usually require special feed treatment.

5.2.5   Resin ion exchange. Resin ion exchange was first used commercially in the processing of uranium during 1952, in South Africa. Strong base resins were used and the early installations were similar to those used in water treatment, the liquids containing uranium passing through columns containing fixed beds of resin until breakthrough. Later, special resins were developed, and a 'moving bed' column ion exchange technique was used in Canada, in 1957. The resin-in-pulp process was first operated commercially in the U.S.A., in 1955, for the treatment of liquors obtained from ores which produced much slimes, making it difficult to obtain a clear liquid. The resin was contained in baskets which were contacted with leach liquors from which the sand, or larger particle size fraction had been removed. A method of suspending resin directly in an agitated pulp, with separation and countercurrent movement of the resin and pulp through multiple stages of a continous operation was developed at the same time. An excellent review of these early processes for uranium extraction is available.[7]

Resin ion exchange suffered from two difficulties, however. The first was poisoning of the resin, particularly by cobalticyanide in South Africa, but a number of other ions built up their concentrations in the resin, lowering its capacity. The second was that column ion exchange is a non-continuous operation and the efforts to develop continuous processes were not wholly successful. Thus, during the mid 1960s, resin ion exchange was progressively replaced by solvent extraction methods, particularly in South Africa, until in 1967 there was only one plant in the U.S.A. using fixed bed and one using moving bed ion exchange, and six using resin-in-pulp methods for specific reasons.[7]

A major problem with solvent extraction however, is the need to produce an aqueous liquor containing less than 10 ppm solids. With the development recently of several systems for achieving continuous (more or less) ion exchange, and in view of the fact that unclarified leach liquors can be treated in them, it seems likely that continuous ion exchange will, at least to some degree, be used in new plants where until recently solvent extraction would have been used.

The contactors which have been developed for continuous ion exchange in general, use upflowing liquor to fluidise a short column of resin within a compartment in a column or series of tanks, and the resin is moved to the adjacent compartment by arresting the upflow of liquid and allowing it to fall, or by pumping it, or by some other means. Eventually the resin in the last compartment leaves the extraction system and passes for elution, before being returned to extract more uranium. The contactors have been described briefly[14] and it was shown that when resin ion exchange and solvent extraction are both technically feasible, and with the assumptions made, solvent extraction is cheaper for liquors containing more than 0.9 g/l U_3O_8, continuous resin ion exchange is cheaper for liquors containing less than 0.35 g/l U_3O_8, and between those two levels of concentration the study could not identify the more economic process. Clarification using a sand filter before solvent extraction was included in the costs for the latter process.

A continuous ion exchange resin contactor, of a design developed in South Africa, came into use in 1977 in that country in a circuit which is recovering uranium from 100 000 tonnes per month of reclaimed solids, including material from the slimes dam.[15] A fluidised bed generated by the upflow of solution in a multistage continuous countercurrent ion exchange column of this type is unrestricted and will allow free passage of suspended particles larger than about 40 μm. The maximum allowable size of an ore particle in the bed is determined by its free-settling velocity, which must be less than that of the smallest resin particle. The resin-in-pulp grade strong base resins at present available have a specific gravity of 1.1 and if a silica slurry is to be treated in a fluidised bed of this resin the separation between sand, not to be fed into the contactor, and slime would have to be made at about 74 μm. In addition, the specific gravity of the slurry must not exceed 1·07. This corresponds to a silica slurry containing 10% w/w suspended solids in a liquid of specific gravity 1·00. Slurries of higher specific gravity moving at practicable flow rates would lift the resin out of the column.

The two constraints, maximum particle size 74 μm and slurry specific gravity not greater than 1·07, can be met by standard hydrocyclones. The sand product discharged from the underflow of these must be washed in a series of hydrocyclone washing stages, but the low capital cost of the hydrocyclones must make their use in conjunction with continuous ion exchange very attractive for the treatment of liquors containing much slimes.

5.3   In situ and percolation leaching

It has been known for many years that certain bacteria which occur in caverns and other voids underground, including mines, oxidise ferrous ions to ferric, and sulphide to sulphate, in order to obtain their energy. They also require oxygen and water, in addition to the usual nutrients used by bacteria. Under suitable conditions, therefore, they will oxidise sulphides of iron present in rocks and form an aqueous solution of ferric sulphate which automatically buffers itself at a pH around 2 because of hydrolysis or redissolution of the basic ferric sulphate thus formed. This ferric solution is a very effective oxidising leaching solution for other reducing minerals, including sulphides containing copper, and minerals containing U(IV), and the ferrous ions produced are re-oxidised by the bacteria. These have usually been known as *Thiobacillus thio-oxidans*, *Th. ferro-oxidans*, *Ferrobacillus ferro-oxidans* and *F. thio-oxidans* although some of these species may be identical. They can become adapted to a remarkable degree so as to withstand quite high concentrations of metal ions, and bacterial leaching has been used for more than 20 years to recover copper and uranium from dumps of waste rock from mining operations and from specially built heaps of low grade ore. The subject has been reviewed recently.[16]

A further development of this kind of leaching process is taking place at Agnew Lake mine, near Sudbury, Ontario. Leaching is being carried out underground in an operation designed to produce eventually 1 M pounds of U_3O_8 per annum. The mine was developed initially during 1968–70 as a conventional underground system; the production shaft was sunk to 3,400 ft. and crosscuts were completed on six levels. The mine was temporarily closed in 1971 because of the low uranium price, and in 1974 a two year pilot test began on the feasibility of recovering the uranium by *in situ* leaching. Development work is now in progress and has been described.[17]

The ore is blasted heavily to cause fragmentation to lumps smaller than 6 inches. This, of course, causes the rock to occupy a larger space than when it was solid and about 30% of the volume of ore is taken to the surface for heap leaching. Leaching underground is carried out on 100–200 ft. lifts, for example between the 1,300 ft. and 1,500 ft. levels. Using the trickle technique a 4 inch diameter perforated pipe is run along the top of a broken ore stope and sulphuric acid at pH 1·5 is run through. This percolates through microscopic cracks in the ore and through the fractures produced by blasting, and is collected at the 1500 ft. sill and pumped to the surface. The stopes are seeded with bacteria to speed up the oxidation of pyrite and pyrrhotite present, so as to achieve an acid ferric sulphate leaching medium. The ore body dips steeply and this has created difficulties in distributing the solution evenly and in future stopes will be plugged with concrete on sump levels to contain the acid, and will be flooded with leach liquor. It will be necessary to maintain a suitable level of oxygen in the solution.

The leach solution, containing about 0·25 g/l of uranium, is not clarified and the uranium is recovered from it using strong base anion exchange resin in Himsley type continuous contactors. The uranium is stripped by backwashing the resin with sulphuric acid, the solution is clarified using a sand filter and passes to a solvent extraction circuit. A tertiary amine is used in three stages of extraction and uranium is stripped from it using ammonium sulphate solution. Ammonium diuranate is precipitated on addition of ammonia.

5.3.1   In situ leaching. A development of even greater significance than underground leaching in mines, at least in the short term, is the introduction of true *in situ* leaching in the U.S.A. and probably in the near future in Australia also. In this method of recovering uranium the ore is not disturbed at all. Wells are drilled at the corners of a square or rectangular grid, about 50 ft. apart, with one well in the centre of each square, thus making two intersecting grids. One grid of wells may be used for injecting leach solution and the other set used as production wells from which the pregnant solution is pumped, or each grid may be permanently used for one purpose.

Certain criteria must obviously be met before an ore body can be worked by using this technique. The uranium must be in porous rock and this should preferably have immediately underneath it a layer of non-porous rock. Also there should not be any badly fractured or channeled structures in the vicinity, although in one of the early patents dealing with *in situ* leaching,[18] cement grouting or the use of special plastics or gels was proposed as a way to seal off regions of possible leakage. In the same patent a leaching technique was described in which the ore body was initially saturated with the leaching solution, carbonate-bicarbonate, at low pressure and flow rate to open up leaching channels. A higher pressure differential and alternate use of the grids of wells for injection and production was then proposed, in order to develop the leaching channels within the ore body and minimise loss of solution outside the area from which it can be recovered.

A later technique[19] used the natural flow of ground water through the ore deposit as a containment shell for leaching solutions but the following requirements had to be met:

1 The ore should occur in a more or less horizontal bed with an impervious layer below it.

2 The ore must be located below the static water table.

3 The direction and velocity of the natural water flow through the ore must be known.

4 The uranium minerals in the ore must be amenable to the proposed leaching process.

5 The ore body must be of sufficient size and grade to justify the cost of the operation.

Generally speaking, this is the basis of the kind of *in situ* leaching process being used at present, although some additional requirements must be added. In a paper[20] presented in 1975 information was given concerning what was described as the only operating commercial sized uranium solution mining facility at that time. This was in Texas, and it was producing 250 000 lb. of U_3O_8 per annum and was planned to produce 1 M lb. per annum after operating experience had been gained. The smaller unit cost \$7 M and the staff consisted of 50 persons at the plant and 12 supporting staff members at the local company office.

The geological formation in which the uranium occurs runs parallel to the Gulf of Mexico and represents flood plain deposits, deltaic deposits or shallow marine lagoonal deposits, and the host rocks are sands and sandstones containing volcanic ash which is generally accepted as being the source of the uranium. The reducing agent which was necessary to precipitate the uranium leached from the ash by the ground water is considered to have been hydrogen sulphide gas derived from deeper-seated accumulations of hydrocarbons. The hydrogen sulphide moved upwards along fault zones into the sand region which was favourable for deposition of uranium.

The sandstone formation is a fresh water aquifer and supplies a number of communities with drinking water, although in the vicinity of the uranium deposit the high level of radioactivity in the water makes it non-potable. Thus it is necessary to ensure that no leach solution leaves the area and that when the area is exhausted the reservoir solution is returned to substantially its original condition. A group of chemical and petroleum engineers, geologists and hydrologists, zoologists, biologists and radiologists was set up to determine the baseline environmental conditions and the monitoring requirements. A baseline study conducted by an independent consultant included bacteria, insects, birds, amphibia and fish, roots and stems, grasses and shrubs, and soils and streams.

All water wells within five miles of the site were sampled and analysed before work was started, the determinations including Ca, Na, Fe, Mo, HCO_3^-, SO_4^{2-}, Cl^-, NO_3^-, NO_2^-, NH_4^+ as well as pH and total dissolved solids. The wells were also analysed for gross α- and gross β-radiation. Whenever the gross α value exceeded 3 pCi/l a barium coprecipitation for radium 226 was performed and a lead 210 determination was performed whenever the gross β value exceeded 50 pCi/l.

Before operations started the same determinations of ions were carried out on water from 28 wells in the area to be leached and also of Mg, F, As, Ba, B, Cd, Cu, Cr, Pb, Mn, Hg, Ni, Se, Ag, U, Zn, simple cyanide and alkalinity.

The injection wells, located on a square grid at intervals of 50 ft. were completed using 4 inch PVC plastic pipe, and the production wells were 6 inches in diameter to accommodate pumps down the holes. All wells had screens opposite the ore zones to permit leach liquor to flow into the zones and to pump pregnant solution out. There were 46 patterns covering approximately 3 acres and the leach field had a ring of monitor wells around it to determine whether leaching solution had migrated away from the leach area. Also, 4 shallow monitor wells were placed above the ore body to detect the presence of leach solutions in shallow aquifers.

If leach solution were to be found in a monitor well, the injection in that area of the field would be decreased while the production was continued. The natural reservoir fluid would then sweep past the monitor well into the leach area, returning the leach solution to the area from which it would be recovered.

After leaching of an area treated by such a process is complete it is necessary to restore it. This is done by pumping out water containing the reagents which have been used to leach out the uranium and continuing to pump out the reservoir fluid passing into the region until substantially all of the reagents have desorbed from the rock. The reagents must be disposed of and it may be possible to pump the solution into a deep well. Since the leaching reagents used are most often ammonium carbonate-bicarbonate or sulphuric acid, and the oxidant is hydrogen peroxide or oxygen injected into the solution under pressure, this seems a reasonable method.

The chemistry of the *in situ* operations is basically the same as that used in conventional uranium production although the use of hydrogen peroxide can involve the formation of gas in the strata. Carbon filters are used to remove silt from the pregnant liquor and they are cleaned by back-washing. Uranium is recovered by ion exchange.

One of the main problems encountered in designing an *in situ* leaching operation for a particular parcel of ore is to try to assess the behaviour of the rock and solution when they are in contact. For example, if a sulphuric acid leach process is to be used acid will be consumed by the host rock, also gypsum is likely to be deposited within it. Iron appearing in the solution may also have to be disposed of. If the carbonate-bicarbonate system is to be used, the sodium salts are more likely to affect clays present and may cause swelling, so that ammonium is usually preferred. Calcium may also deposit as carbonate within the rock.

It is usual to carry out simulation studies to gain an idea of what may happen in practice. A paper describing one way of doing this was presented in 1978.[21] This involved packing particulate ore, from which the smaller size fractions had been removed, into a vertical column, transferring this to a horizontal position and passing solution through the column. Clearly, it is very doubtful whether what happens in it reflects with any degree of accuracy what would happen were the same solution to pass through the rock which had not been disturbed. Therefore, it is now the practice to use a core sample enclosed in a tube. Early tests showed that the uranium present in such a core was much easier to leach out than was the case in practice. The difference was found to be due to oxidation of the uranium taking place between drilling the core and carrying out the tests. Now core samples are at once sealed in an inert atmosphere and preferably stored in a freezing chamber at about $-40°C$. Because such rapid changes are occurring in the technology of *in situ* leaching on all fronts, details of other recent publications in the field are not given here.

5.4   Extraction from phosphoric acid

About 80 percent of the world production of phosphate rock is from deposits of marine phosphorite. These are believed to have been formed by divergent upwelling of sea water, by deposition in warm currents along coasts or by deposition from water on stable shelves. The richest of the marine deposits have been concentrated or enriched by secondary processes, by submarine re-working or by leaching during weathering. The phosphate rock also contains carbonaceous material capable of reducing uranium to U(IV) and, in consequence, the uranium present in the sea (about 3ppb U) or in ground water in contact with the rock will tend to deposit on it. Thus the phosphate rock in Florida and the western U.S.A. contains between about 50 and 200 ppm uranium. When this is used to produce wet-process phosphoric acid the uranium reports in the 27–30% P_2O_5 phosphoric acid stream and methods are now being developed to recover it.

There are a number of problems to be overcome:

(i)   organic material from the phosphate rock remains in the acid and if solvent extraction is used it produces very troublesome emulsions;

(ii)   since phosphoric acid forms complexes with many metals it is difficult to predict the behaviour of these in the 30% acid;

(iii)   if uranium is extracted in cationic form the extractant must be an extremely powerful one so that stripping in the conventional manner will also require strong acid;

(iv)   a powerful extractant will remove other metal ions present in the phosphoric acid and these will have to be stripping either with the uranium or separately.

A solvent extraction process for recovering the uranium from phosphoric acid produced from Florida phosphate rock was developed in the early 1950's but was used commercially for only a short period because of a number of troublesome problems in the process, and the discovery of adequate reserves of uranium ore.

Since that time the production of phosphate fertilisers has greatly increased and the amount of U_3O_8 dissolved in wet-process phosphoric acid from the Florida rock is now probably around 2,000 tons per annum. In 1967 a programme was started at the Oak Ridge National Laboratory in the USA to develop an improved process to recover this uranium and a large number of potential extractants were screened. The usual extractants for uranium, tributylphosphate and long-chain tertiary amines, did not have sufficient extracting power in the phosphoric acid system and the synergistic extractant combination of di(2-ethylhexyl) phosphoric acid (D2EHPA) plus trioctylphosphine oxide (TOPO) in an aliphatic diluent of high boiling point showed the most promise in 1972.

A process was designed and tested in a laboratory scale continuous multi-stage mixer-settler unit.[22] The synergistic extractant behaves efficiently only with U(VI) whereas most of the 0·1–0·2 g/l of uranium in the phosphoric acid is U(IV). This is oxidised by adding sufficient oxidant, sodium chlorate, hydrogen peroxide or oxygen, to oxidise all of the iron present to Fe(III), after cooling the acid to 40–45°C. The uranium is then extracted with 0·5 M D2EHPA–0·125 M TOPO and is stripped from the solvent by contacting it with a phosphoric acid solution (recycled raffinate) containing ferrous sulphate, which reduces the uranium to U(IV) to give an aqueous solution containing about 12 g/l uranium. This is then treated in a second cycle of solvent extraction. After re-oxidation the uranium is extracted into 0·3 M D2EHPA–0·075 M TOPO and at least half of the raffinate is recycled for further stripping, the rest being returned to the first cycle extraction step. The organic phase is washed with water to remove extracted phosphoric acid and then stripped with 2–3 M ammonium carbonate solution, which cause precipitation of ammonium uranyl tricarbonate.

The same ingenious method of stripping by changing the valency of uranium was used in a variation of the process.[23] A commercial mixture of mono- and dioctylphenylphosphoric acids (OPPA) is used to extract the uranium from the phosphoric acid in the first cycle. It is much less expensive and has a higher extracting power than 0·5 M D2EHPA– 0·125 M TOPO, and it extracts U(IV) rather than U(VI). In the phosphoric acid all of the uranium will be in this valency state if the concentration of ferrous iron in the solution is 0·5 g/l or higher. This is generally the case.

In the first cycle of the latter process the phosphoric acid is cooled to 40–45°C and the uranium is extracted with 0·3–0·4 M OPPA in an aliphatic diluent. Uranium is stripped from the organic phase with 10 M phosphoric acid containing sodium chlorate which oxidises the uranium, to give a solution containing 15–20 g/l uranium. This solution is diluted to 6 M phosphoric acid and the uranium is extracted in the second cycle, again in three stages, with 0·3 M D2EHPA–0·075 M TOPO. The organic phase is scrubbed with water and stripped with 2–3 M ammonium carbonate solution.

OPPA and the pyrophosphates which can be used as extractants for uranium also extract ferric iron whereas the D2EHPA-TOPO mixture does not extract the iron well, but does extract rare earths, vanadium and group VI metals. Since any iron present in the organic phase in the second cycle precipitates as an impurity with the uranium in stripping, the use of the more expensive synergistic extractant in the second cycle is necessary.

The formation of emulsions by the organic material (humic acids) in the phosphoric acid can be minimised by contacting the phosphoric acid under controlled conditions with an organic solvent to produce a small amount of emulsion containing more than 99% of the organic material plus some gypsum. This emulsion is dealt with in a side stream of small volume to recover the solvent and separate the organic material. Despite the removal of some gypsum by this means, deposits of this material are still troublesome.

The principal choice of materials for tanks and pipelines for the recovery of uranium from phosphoric acid has been fibreglass-reinforced plastics. This is associated with the fact that very stringent limits are placed on the quantity of organic phase which remains in the phosphoric acid raffinate. It can cause very serious problems with materials of construction in the downstream phosphoric acid plant.

6   References

1 Associated Scientific and Technical Societies of South Africa. *Uranium in South Africa 1946–1956*, 2 vols. Johannesburg: The Association of Scientific and Technical Societies of South Africa, 1957.

2 Runnalls O.J.C. Department of Energy, Mines and Resources, Ottawa. The uranium industry in Canada. A brief submitted to the Cluff Lake Board of Inquiry, Regina, Saskatchewan, May 1977. Provided at a Short Course on the Extractive Metallurgy of Uranium, University of Toronto, May 1978.

3 Davis M. European needs for uranium in relation to world supply and demand. In *Geology, Mining and Extractive Processing of Uranium*, Jones M.J. ed. London: Institution of Mining and Metallurgy, 1977, 1–7.

4 Towards a new energy strategy for the European Community. COM (74) 550 final. Supplement Bull. Europ. Commun., no. 4, 1974, 35p.

5 Dayton S. Uranium. *Engineering and Mining Journal*, 1978, **179** (11), 57–59.

6 The Uranium Institute. *Uranium supply and demand*. London: Mining Journal Books Ltd., 1977.

7 Merritt R.C. The extractive metallurgy of uranium. U.S.A.: Colorado School of Mines Research Institute, 1971.

8 Langmuir D. Uranium solution-mineral equilibria at low temperatures with applications to sedimentary ore deposits. *Geochimica et Cosmochimica Acta*, 1978, **42**, 547–569.

9 Needes C.R.S., Nicol M.J. and Finkelstein N.P. Electrochemical model for the leaching of uranium dioxide: 2 alkaline carbonate media. In *Leaching and Reduction in Hydrometallurgy*, Burkin A.R. ed. London: Institution of Mining and Metallurgy, 1975, 12–19.

10 Nicol M.J., Needes C.R.S. and Finkelstein N.P. Electrochemical model for the leaching of uranium dioxide: 1 acid media. As reference 9, pp. 1–11.

11 Burkin A.R. and Monhemius A.J. Acid leaching of uranium ores using hydrogen peroxide and Caro's acid. Paper presented at the CIM conference of Metallurgists, Montreal, Canada, August 1978.

12 Smith S.E., Lapage R. and Garrett K.H. British Patent 1 328 242, 1973.

13 Forward F.A. and Halpern J. U.S. Patent 2 797 977, 1957.

14 Brown A.J. and Haydon B.C. A comparison of liquid and resin ion exchange processes for purification and concentration of uraniferous solutions. Paper presented at the CIM conference of Metallurgists, Montreal, Canada, August 1978.

15 Anon. NIMCIX is a commercial success at Blyvooruitzicht. *Engineering and Mining Journal*, 1978, **179** (11), 178–181.

16 Tuovinen O.H. and Kelly D.P. Use of micro-organisms for the recovery of metals. *International Metallurgical Reviews*, 1974, **19**, 21–31.

17 Thomas R.A. Agnew Lake Mines: Taking giant steps in solution mining. *Engineering and Mining Journal*, 1978, **179** (11), 158–161.

18 Livingston C.W. U.S. Patent 2 818 240, 1957.

19 Anderson J.S. and Ritchie M.I. Solution mining of uranium. *Mining Congress Journal*, 1968, **54**, 20–23, 26.

20 Davis G.R., Miller R.E. and Swift G.G. *In situ* leach mining for uranium. Uranium ore processing. Proceedings of an advisory group meeting organised by the International Atomic Energy Agency and held in Washington, DC, 24–26 November 1975. Vienna: International Atomic Energy Agency, 1976, 193–202.

21 Sundar P.S. *In situ* leaching simulation studies of uranium ores. Paper presented at the AIME annual meeting, Denver, Colorado, February 1978.

22 Hurst F.J., Crouse D.J. and Brown K.B. Recovery of uranium from wet-process phosphoric acid. *Ind. Eng. Chem. Process Des. Develop.*, 1972, **11** (1), 122–128.

23 Hurst F.J. and Crouse D.J. Recovery of uranium from wet-process phosphoric acid by extraction with octylphenylphosphoric acid. *Ind. Eng. Chem. Process Des. Develop.*, 1974, **13** (3), 286–291.

The electrolytic production of zinc

A. J. Monhemius

1　　　Introduction

The modern extractive metallurgy of zinc is dominated by the electrolytic process. This elegantly simple roast-leach-electrowin process now accounts for almost 80% of the production of primary zinc in the world. The process was first used on a commercial basis in 1916 and its use grew steadily until, by the mid-sixties, it accounted for about 50% of zinc production. Its growth since then, compared with the other methods of zinc production, which are all pyrometallurgical processes, is shown in Table 1.

Table 1. Zinc production processes[1]

Process	Percentage of total production capacity			
	1965	1970	1975	1980*
Electrolytic	49	57	70	79
Horizontal retort	26	14	3	1
Vertical retort	11	11	8	5
Electrothermic	8	7	7	6
Imperial smelting	6	11	12	9

* estimate.

Looking first at the pyrometallurgical processes, it is obvious that the old Horizontal Retort process is virtually extinct. The other two older processes, the Vertical Retort and the Electrothermic process, are both in steady decline, the former somewhat more rapidly than the latter. Even the Imperial Smelting Process, which is a modern process, introduced only in 1959, and possessing unique advantages such as the ability to treat lead-zinc and other complex raw materials, appears to have passed its peak in its share of total zinc production capacity.

In contrast with the pyrometallurgical processes, the percentage of zinc produced by the electrolytic process has seen a sharp increase over the past 15 years or so, and one of the purposes of this review is to discuss some of the reasons for this growth. However, in common with the rest of non-ferrous extractive metallurgy, the technology of primary zinc production is not static. In spite of its present dominating position and advanced state of technological development, the modern electrolytic process can undergo further improvements and modifications which will make it capable of treating a wider variety of feed materials. There are a number of processes, in various stages of development, which fall into the category of improvements or modifications of the basic electrolytic process and the more important of these are included in this review. Some of these processes involve radical change to parts of the basic process, but the common theme is that they all aim to produce pure zinc sulphate solutions, from which zinc metal can be electro-won using the technology developed over the six decades of existence of the electrolytic process.

2   The modern electrolytic process

In order to appreciate the reasons why the electrolytic route has grown to its position of overwhelming dominance in the primary production of zinc, it is necessary first to examine briefly the basic principles of the process. A flowsheet of the basic process is shown in Fig. 1.

Zinc sulphide flotation concentrate is roasted in air to form calcine, which is primarily zinc oxide. Sulphur in the feed is oxidised to sulphur dioxide, which reports in the roaster off-gases and is used to make by-product sulphuric acid. The calcine passes to the leaching plant, where it is dissolved in spent electrolyte, returned from electrowinning and typically containing 50 g/l Zn and 150–200 g/l H_2SO_4. Leaching is usually carried out in a two-stage counter-current system, in which the calcine enters the second or 'neutral leach' stage. Here it is partially dissolved in the weakly acidic liquor coming from the first or 'acid leach' stage. It is important that the liquor leaving the neutral stage is neutralised to a pH of about 5. Under these conditions, any iron in solution is precipitated during leaching as ferric hydroxide, which has a vital scavenging role and removes, by adsorption or co-precipitation, many deleterious impurities, such as As, Sb, and Ge, from the zinc sulphate solution. The solid residue from the neutral leach stage, containing

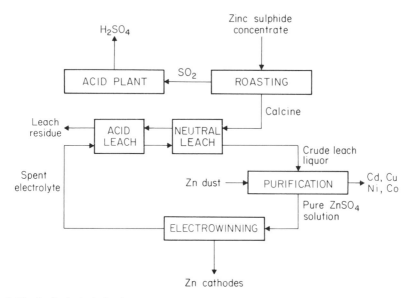

Fig. 1. The basic electrolytic zinc process.

undissolved calcine, passes to the acid leach stage, where it is contacted with the incoming spent electrolyte and dissolution of the calcine is completed. The liquor from the neutral leach stage will contain about 150–200 g/l Zn plus minor amounts of other metal impurities, which must be removed prior to electrolysis. Metallic impurities comprise mainly Cd, Cu, Ni and Co. These are removed from solution by cementation with zinc dust, usually in two stages, although in some plants a separate operation for cobalt removal is also required.

Following purification, the strong zinc sulphate solution is passed to the tankhouse, where zinc is electrowon on to aluminium starter sheets, using inert lead anodes. Thermodynamically, the electrowinning of zinc from aqueous solution is not possible, since the reduction potential of zinc lies well below that of water. It is only the high overpotential for the generation of hydrogen at a zinc surface that makes the process possible. The hydrogen over-potential depends on a number of process variables, but the most important of these is the impurity content of the electrolyte. The presence of certain impurities, at concentrations as low as one part in 10^7, can cause a drastic lowering in the hydrogen overpotential and thus entirely stop the electrolysis of zinc. It is the requirement for extreme purity in the zinc sulphate electrolyte which is the primary constraint on all the prior steps in the process.

Improvements have been made in all parts of the process over the years and, although the basic process principles remain the same, the modern electrolytic plant bears little resemblance to the first generation plants, constructed during the

First World War. The development of the electrolytic process has been comprehensively reviewed recently[6] and so only a brief outline of modern practice is included here.

Modern concentrate roasting is done in fluid-bed roasters. Roasting is autogenous in these efficient, highly automated units and so there is no external fuel requirement. The off-gases are rich in SO_2, resulting in a high degree of conversion to sulphuric acid in the acid plant and low SO_2 emissions in the exhaust gases. Heat in the form of process steam is also recovered from the furnace off-gases in waste heat boilers.

It is in the leaching plant that the greatest improvements in the overall process have taken place. This has happened particularly over the past fifteen years or so and it is primarily these improvements which account for the sharp rise in the use of the electrolytic process over this period. The developments in leaching practice will be discussed more fully in the next section.

Methods of purification of the leach liquors have changed little over the years. The two major steps are still the precipitation of ferric hydroxide, usually incorporated in the leaching process, followed by cementation with zinc dust to remove other metallic impurities from solution. However, even here, improvements have been made, particularly in the cementation step. At least three purification flowsheets are in current use, all using additions of catalysts to the cementation step, to improve the usage of zinc dust and thus to reduce the amount of zinc recycle in this part of the process. One purification flowsheet involves the use of arsenic and copper as cementation catalysts and the other two use antimony.[2]

In the tankhouse, major improvements have been confined to the mechanisation and automation of the cathode handling operations. This has dramatically lowered tankhouse labour requirements and has also enabled larger cathodes to be used, thus reducing the number of cells, and therefore tankhouse size required for a given production capacity.

Hence the modern electrolytic zinc plant is a highly efficient, technically refined, production unit. It is capable of a high degree of automation and can incorporate modern methods of process instrumentation and control. Consequently, overall labour requirements are low. There is, at the present time, no foreseeable shortage of the principal raw material for the process, i.e. high quality zinc sulphide concentrates, at least to the end of this century.[1] Unless there is a dramatic shortfall in the availability of electrical power, which is the principal energy requirement of the process, it seems likely that the electrolytic route will retain its dominant position in the production of primary zinc metal.

Nevertheless, no process is perfect and the zinc electrolytic process is no exception. It is perhaps ironical that the part of the process where the greatest improvements to the overall metal recovery efficiency have already been made, namely the leaching operations, is the part which requires still more improvements and modifications and which is the object of a considerable amount of current research and development effort. This work has basically two objectives. Firstly, efforts are being made to modify the existing process to produce more environmentally acceptable effluents, in particular the solid residues generated in leaching. This has led in certain cases to proposals for radically different methods of leaching, for example direct leaching of zinc sulphide concentrates, using techniques such as pressure leaching. Secondly, the range of raw materials treatable by the electrolytic process is being extended beyond high-grade zinc sulphide concentrates to include such materials as zinc silicate ores and complex sulphide ores, containing zinc together with large quantities of other metals such as iron, copper and lead. Again, it is primarily modifications to leaching methods which allow incorporation of these feed materials into the overall process, although new solution purification procedures may also be required, particularly where complex ores are to be treated.

Thus the major part of this review is concerned with developments that have already taken place, and those that are likely to do so, in the leaching of zinc-bearing materials.

3   Zinc leaching practice

Virtually all zinc sulphide concentrates contain iron at concentrations typically falling in the range 5–12%. During roasting, particularly under the conditions used in modern fluidised-bed roasters, most of this iron combines with zinc to form ferrites. These are complex oxides of variable composition, in which iron is in solid solution with zinc, but they can be represented as $ZnFe_2O_4$.

Zinc ferrites are insoluble under the conditions of temperature and acidity used to dissolve the major zinc oxide portion of the roaster calcine. Thus in the traditional two-stage leaching process, ferritic zinc reported in the solid leach residues, together with nearly all the lead, silver, gold and gangue materials contained in the original feed. This loss of zinc in the leach residues resulted in overall zinc recoveries of only 85 to 93% and this was the most serious drawback of the electrolytic process, particularly where high-iron feeds were involved.

It was known that zinc ferrites would dissolve readily in sulphuric acid, provided strong acid solutions were used and the temperature was maintained close to the boiling point. However these conditions also resulted in the dissolution of most of the iron in the residues as well as the zinc. This iron had to be removed before the zinc solution could be returned to the main leach circuit. The removal of large quantities of iron from solution was not technically possible until about 15 years ago, because the only known method of iron removal was precipitation as ferric hydroxide. This is a gelatinous material which occludes a great deal of solution and the difficulties of filtering and washing material which contains large quantities of ferric hydroxide are virtually insuperable.

The situation was dramatically changed, however, by the development of methods whereby iron could be precipitated as crystalline, easily filterable materials, which could be readily separated from solution. Such materials include the basic iron sulphates known as jarosites, and also iron oxides, both hydrated and anhydrous. This development enabled existing leaching practices to be modified, so that not only oxide zinc but also ferritic zinc could be recovered from the calcine feed.

A number of methods for the removal of iron from solution have been introduced – the Jarosite process and its variant, the Conversion process; the Goethite process; and the Hematite process. Over the past decade or so, many existing electrolytic zinc plants and all new plants have incorporated one or other of these processes into their leaching circuits, with the result that typical overall recoveries of zinc in the electrolytic process have risen to 95–97%. Each of these various iron-removal processes is described in the discussion which follows.

3.1   The Jarosite process

In the Jarosite process, iron is precipitated from acidic sulphate solutions as one of a group of basic ferric sulphates known as jarosites. The composition of jarosites is $MFe_3(SO_4)_2(OH)_6$, where M represents a monovalent cation from the group comprising Na^+, NH_4^+, K^+, Ag^+, Rb^+, $\frac{1}{2}Pb^{2+}$ and H_3O^+. Precipitation is brought about by adjusting the solution pH to about 1·5, at a temperature of about 95°C and adding a source of the monovalent cation, NH_4^+ or Na^+ being the most usual cations used industrially. The reaction which occurs may be represented, in simplified form, as follows:

$$3Fe^{3+} + 2SO_4^{2-} + M^+ + 6H_2O \rightarrow MFe_3(SO_4)_2(OH)_6 + 6H^+ \tag{1}$$

It may be seen that in common with other hydrolysis reactions, hydrogen ions are produced and these have to be neutralised to maintain the pH at the required value during precipitation. The jarosite precipitates from solution in a crystalline form which is easily separated from solution.

Very similar versions of this process were developed independently, and patent applications lodged within months of each other in the mid-sixties, by three zinc companies: Asturiana de Zinc S.A.;[3] Electrolytic Zinc Company of Australasia Ltd.;[4] and Det Norske Zinkkompani A/S.[5] (now Norzink A.S.). These companies have since combined for the licencing of the process. The Jarosite process in various modifications is the most widely used method for iron removal from zinc leach liquors. In a recent paper by Gordon,[6] out of 17 plants quoted as using ferrite leaching, 13 of these use the Jarosite process for subsequent iron removal.

The Jarosite process has been described quite extensively in the literature.[7–11] It is very flexible, capable of being both easily integrated into existing leach plants and also adapted to cope with different plant practices. An excellent discussion of the options available within the overall process design may be found in the paper by Gordon and Pickering.[11] For the purposes of this review, the discussion will be confined to the so-called Integrated Jarosite process, used by Norzinc,[7] as this design incorporates most of the various process options. A simplified flowsheet of the process is shown in Fig. 2.

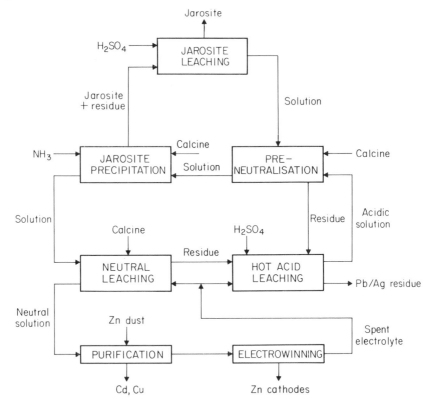

Fig. 2. The Integrated Jarosite process.

The bulk of the calcine enters the plant into the neutral leach reactors, where about 80% of the soluble zinc is dissolved in a three to one mixture of return solution from jarosite precipitation and spent electrolyte from electrowinning. Oxidising conditions are created by additions of manganese dioxide and the small amount of iron present in solution is oxidised and precipitated as ferric hydroxide, carrying with it certain harmful impurities, in particular As, Sb and Ge, which if left in solution would detrimentally affect the electrowinning stage. The neutral leach liquor, which has a pH of about 5 and a zinc content of about 180 g/l, is separated from the solid residues and passes to the purification stages prior to electrowinning.

The solid residues from neutral leaching are transferred to the hot acid leach tanks where they are leached at 95°C with spent electrolyte and make-up sulphuric acid. Under these conditions, the zinc ferrites break down and the majority of zinc and iron is leached out, together with much of the copper and cadmium in the residues. The solid residues from hot acid leaching contain all the lead and silver present in the original calcine and, in many plants, this material is a valuable by-product of the process, which can be sold as feed to a lead smelter.

The solution leaving hot acid leaching has a residual acidity of about 40 g/l H_2SO_4. This is partially neutralised to about 10 g/l by addition of calcine in the pre-neutralisation tanks. Only the oxide zinc in this calcine dissolved and so the ferritic residue is returned to hot acid leaching.

The pre-neutralised solution is then ready for jarosite precipitation. This is carried out at 95°C at a pH controlled at 1·5. Ammonia is added to precipitate iron as ammonium jarosite, $NH_4Fe_3(SO_4)_2(OH)_6$, and calcine is added to neutralise the acid produced by the reaction (Equation 1). Precipitation is continued until the iron concentration has been lowered from about 20 g/l to about 1 g/l. This amount of iron is deliberately left in the solution which is then sent to neutral leaching, where the iron concentration drops to $< 0·01$ g/l due to ferric hydroxide precipitation, the benefits accruing from this having already been mentioned.

The jarosite residue contains the zinc ferrites arising from the calcine added as a neutralising agent during jarosite precipitation. In order to recover this zinc, the final step in the process is an acid wash of the jarosite residue under conditions similar to those used in the hot acid leach section. This releases ferritic zinc into solution, while the jarosite remains virtually unattacked. The solution from jarosite leaching is returned to pre-neutralisation and the acid-washed jarosite is discarded.

The overall recovery of zinc by this process, from calcines containing 56–57% Zn and 10–11% Fe, is 98%. Recoveries of other valuable metals contained in the calcine are as follows: Cd, 97; Cu, 90; Pb, 82; Ag, 82%.

The main advantages of the Jarosite Process, compared with the other iron-removal processes discussed subsequently, arise principally from the nature of jarosite itself. Being a basic sulphate, it is precipitated from relatively acid solutions, thus allowing efficient usage of calcine added as neutralising agent. It also enables excess sulphate to be removed from the closed leaching cycle. This is a distinct advantage in many plants where excess sulphate originating from the calcine creates sulphate balance problems. Furthermore once formed, jarosite is very resistant to acid attack and therefore undissolved calcine mixed with jarosite can be readily recovered by acid washing. These properties confer a considerable degree of flexibility on the operation of the process and allow for optimisation to meet particular process requirements.

The principal disadvantages appear to be two-fold. Firstly the necessity to add a precipitation reagent, normally ammonia, to cause the formation of jarosite. The theoretical consumption of ammonia is 37 kg/1,000 kg of Jarosite, but in practice, the consumption is usually less than 30 kg/1,000 kg, because ammonium jarosite always contains some H_3O^+ in solid solution. Secondly, the low iron content of jarosite, theoretically 35%, means that considerable quantities are produced for disposal. For a calcine of the composition quoted above, i.e. 56–57% Zn and 10–11% Fe, of the order of 0·53 ton of jarosite will be produced per ton of electrowon zinc.

3.2   The Conversion process

The Conversion process, developed by Outokumpu, is a modification of the Jarosite process in which zinc ferrite leaching and jarosite precipitation take place simultaneously in the same reactor.[12,13] The overall reaction may be represented in simplified form as follows:

$$3ZnFe_2O_4 + 6H_2SO_4 + (NH_4)_2SO_4 \rightarrow 2NH_4Fe_3(SO_4)_2(OH)_6 + 3\,ZnSO_4 \qquad (2)$$

The process depends upon the fact that zinc ferrites are more soluble in sulphuric acid than jarosites. Thus, by controlling the acidity at an optimum value, conditions can be created where ferrites will dissolve while jarosites will precipitate. As shown by the above equation, the reaction consumes sulphuric acid and so the process may be controlled by the feed rate of acid to the Conversion reactor. This is the key difference between the Conversion process and the Jarosite process. In the latter, the starting solution contains excess acid, which is neutralised by the addition of calcine, and there is the danger that improper control can lead to over-neutralisation, which in turn can lead to precipitates with difficult solid–liquid separation characteristics. In contrast, it is claimed that because the Conversion reaction is controlled by adding acid, the problems of over-neutralisation are eliminated and the process is very stable in operation, producing very consistent precipitates. Little detail is available on the operation of this process, but it is believed that the major drawback is that long residence times in the Conversion reactor are necessary.

3.3   The Goethite process.

In the Goethite process, iron is precipitated from solution as hydrated ferric oxide, FeOOH. The process used commercially was developed by the Société de la Vieille Montagne[14] and involves reduction of iron to the ferrous state, followed by oxidation with air at a temperature of about 90°C and at a pH controlled at about 3·0. The reaction involved is:

$$4Fe^{2+} + O_2 + 6H_2O = 4FeOOH + 8H^+ \qquad (3)$$

The oxidation of ferrous iron by oxygen is catalysed by copper present in the leach liquor, and it is vital that the rate of oxidation is balanced by the rate of precipitation so that the concentration of ferric iron in solution does not rise above 1 g/l. There is no reagent requirement to supply monovalent cations, as in the Jarosite process and, theoretically, no sulphate is removed from the process stream in the iron product. However, in practice, sulphate contamination is quite heavy, due to adsorption and the formation of some basic sulphates and the iron product usually contains 2–5 wt % sulphur.

The main differences between the Goethite and Jarosite processes occur after the hot acid leaching of the zinc ferrite residues. In the Goethite process, the liquor from hot acid leaching, containing 100 g/l Zn, 25–30 g/l Fe^{3+} and 50–60 g/l H_2SO_4, is first subjected to a reduction stage, where ferric iron is reduced to the ferrous state by reaction with unroasted zinc sulphide concentrate at 90°C:

$$2Fe^{3+} + ZnS \rightarrow 2Fe^{2+} + Zn^{2+} + S^o \tag{4}$$

After reduction of iron is completed, unreacted zinc sulphide, together with the elemental sulphur formed by the reaction, is separated and returned to the roaster. The solution is then pre-neutralised to 3–5 g/l H_2SO_4 with calcine. The ferritic residue from pre-neutralisation is separated and returned to hot acid leaching and the solution is passed to the precipitation reactor. Air is injected to oxidise the ferrous iron, which hydrolyses and precipitates as crystalline goethite. Calcine is added during precipitation to consume the acid produced by hydrolysis and thus to control the pH at the desired value. Following iron precipitation, a solid–liquid separation is made, with the solution returning to neutral leaching and the goethite precipitate, plus undissolved calcine, being discarded. It is not possible to use an acid wash to recover undissolved zinc in the discard material because goethite, unlike jarosite, would redissolve. A flowsheet of the Goethite Process is shown in Fig. 3. A detailed description of Metallurgie Hoboken-Overpelt's zinc electro-winning plant, which uses Vieille Montagne technology, including the Goethite process, is given in a paper by Van Den Neste.[15]

A rather similar process, which has not yet been used commercially, was developed by the Electrolytic Zinc Company of Australasia Ltd.[16] In this process, iron in the ferric state is precipitated directly, without prior reduction, as a hydrated ferric oxide of undetermined nature. This is done by controlled addition of the iron-bearing liquor to a continuous precipitation reactor, where the ferric iron concentration is maintained at less than 1 g/l. Precipitation is carried out at 70–90°C and the pH is maintained at 2·8 by the addition of calcine as neutralising agent. The reaction involved is:

$$2Fe^{3+} + 4H_2O \rightarrow Fe_2O_3.H_2O + 6H^+ \tag{5}$$

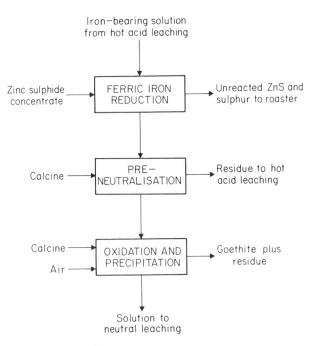

Fig. 3. The Vieille Montagne Goethite process.

Comparison of this reaction with that given above for the Vieille Montagne process shows that more acid is produced by this process and therefore the neutralisation requirements are greater. Consequently, the loss of zinc, due to undissolved calcine in the iron product, is likely to be greater in this process.

The Goethite process does not seem to have the inherent flexibility of the Jarosite process. Very careful control of the conditions during precipitation, especially pH, is required for successful operation of the process. The relative solubility of Goethite in sulphuric acid is a major disadvantage, because the iron residue cannot be acid-washed to recover undissolved zinc arising from the calcine added for pH control. In the new Hoboken plant,[15] this problem is overcome by using carefully selected calcine with a low ferrite content for neutralisation during Goethite precipitation. This type of solution is, however, unlikely to be applicable to plants tied to single major sources of raw material, where gross variations in the compositions of the calcines will not occur. In such plants, the use of the Goethite process would almost inevitably lead to lower overall zinc recoveries than could be achieved with the Jarosite process. This is probably the principal reason for the much wider usage of the latter process.

3.4   The Hematite process

The precipitation of iron as hematite, Fe_2O_3, from residue leach liquors is practised by the Akita Zinc Company in Japan.[17]

Ferritic residue from the main leach circuit is re-leached with spent electrolyte and make-up acid in the presence of SO_2. The reaction is carried out at 95–100°C in autoclaves, lined with lead and acid-resistant brick, at a total pressure of 0·2MPa (30 psi). The errites dissolve readily in the presence of SO_2 and iron enters solution in the divalent state:

$$ZnFe_2O_4 + SO_2 + 2H_2SO_4 \rightarrow ZnSO_4 + 2FeSO_4 + 2H_2O \tag{6}$$

Excess SO_2 is stripped from the solution and copper is removed as sulphide by H_2S precipitation. The solution, containing about 90 g/l Zn, 60 g/l Fe and 20 g/l H_2SO_4 is neutralised with limestone, first at pH 2 to produce a marketable grade of gypsum and then at pH 4·5, with air oxidation to oxidise and precipitate some of the iron and other impurities. The precipitation of gypsum helps to maintain the sulphate balance in the process by removing sulphate formed during the oxidation of SO_2.

The neutralised solution, which contains about 45 g/l iron is then passed to titanium-clad autoclaves, where the iron is oxidised and precipitated as hematite by oxidation with oxygen at 200°C and a total pressure of 2MPa (300 psi). The reaction involved is:

$$2FeSO_4 + 2H_2O + \tfrac{1}{2}O_2 \rightarrow Fe_2O_3 + 2H_2SO_4 \tag{7}$$

The residence time in the autoclaves is about 3 hours. The final solution, after hematite precipitation, contains 3–4 g/l Fe and this is returned to the main leach circuit. A flowsheet of the process is shown in Fig. 4.

At the high temperatures used for iron precipitation in this process, hematite will continue to form even in relatively acidic conditions. Thus the necessity to add calcine to consume protons produced by the hydrolysis reaction is eliminated. This is a major advantage of this process and it means that theoretically no zinc should be lost with the iron residue. However, in practice, the iron residues contain 0·5–1·0% Zn, together with about 3% S,[6] the sulphur arising presumably from the co-precipitation of basic iron sulphates. A second advantage of the process is the high iron content of the hematite residues, theoretically 70% Fe, but in practice closer to 60% Fe, which leads to much smaller quantities of iron residues for disposal. For example, a given quantity of iron precipitated as hematite will have less than half the weight that the same quantity of iron would have if precipitated as jarosite.

The major disadvantage of this process is, of course, the high cost of the pressure equipment used and, to date, the process is only in use in the one Japanese plant.

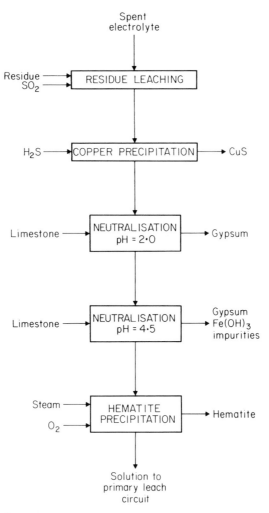

Fig. 4. The Hematite process.

3.5   Future developments

Although the hydrolysis processes described above are technically very successful and acceptable, environmentally they are less so. Jarosite, goethite and hematite, as produced by the respective processes, all contain appreciable amounts of other elements, in particular zinc and sulphur. The zinc arises mainly from undissolved zinc calcine in the iron residues and the sulphur from sulphate, which is either incorporated into the crystal lattice or adsorbed in the precipitate. Gordon[6] has given some data on the amounts of iron residue produced by each process and their approximate compositions. These data are reproduced in Table 2.

Table 2. Compositions and quantities of iron residues[6]

Process	Jarosite	Goethite	Hematite
Wt/wt. concentrate	0·4	0·25	0·18
Fe wt. %	25–28	40–45	58–60
Zn wt. %	4–6	5–8	0·5–1·0
Total S wt. %	10–12	2·5–5	3

As a result of the impurity content none of these materials is suitable for iron making and therefore they have to be disposed of by dumping. The amount of soluble impurities in the iron residues makes environmentally safe disposal a difficult task[18,19] and increasing concern about these problems is being expressed, particularly by European zinc manufacturers.

An alternative way of removing iron from zinc leach liquors, which resulted in an iron product suitable for iron-making, would be an attractive solution to these problems. One method, which in concept, at least, shows promise in this direction, is iron removal by solvent extraction.

Ferric iron can be extracted at low pH's by liquid cation exchangers, such as alkyl phosphoric acids and carboxylic acids. In order to avoid having to neutralise the hydrogen ion produced by the cation exchange reaction, it is preferable to use the method of exchange extraction using the zinc salt of the organic acid:

$$2Fe^{3+} + 3\overline{R_2Zn} \rightarrow 3Zn^{2+} + 2\overline{R_3Fe} \tag{8}$$

where a bar above a species indicates that it is in the organic phase.

Patents on the use of the zinc salts of Versatic acids, synthetic tertiary carboxylic acids, for the extraction of iron from zinc liquors have been granted to Thorsen[20] and to Shell,[21] and details of a laboratory-scale investigation have been published by Van der Zeeuw.[22] Iron is very readily extracted from acidic zinc liquors by this method and the amount of zinc transferred from the organic to the aqueous phase is equivalent to the initial amount of iron plus the initial free acid in the feed liquor. The zinc salt of the carboxylic acid can be easily prepared by direct dissolution of zinc oxide or zinc calcine into the organic phase:

$$2\overline{RH} + ZnO_{(s)} \rightarrow \overline{R_2Zn} + H_2O \tag{9}$$

The problem with this process lies in the stripping and recovery of iron from the loaded organic phase. The method suggested by Van der Zeeuw involves stripping with 6N HCl to form a ferric chloride solution, which could then be pyrohydrolysed to convert the iron to Fe_2O_3 and to recover HCl for recycle.

However an alternative approach to iron stripping is revealed in a recent paper by Thorsen and Monhemius[23] in which it is shown that iron can be directly precipitated from the organic phase as iron oxide by a process called hydrolytic stripping. An iron-loaded carboxylic acid is heated to 150–200°C in the presence of water and the following reaction occurs:

$$2\overline{R_3Fe} + 3H_2O \rightarrow Fe_2O_{3(s)} + 6\overline{RH} \tag{10}$$

The reaction can be rapid. For example at 185°C, total precipitation of iron occurred within 20 minutes. Versatic 10, the acid used in the investigation, was found to be thermally stable at these temperatures and could be recycled to extraction, following hydrolytic stripping. The advantage of this process is that, because the iron is precipitated from an organic phase, there is no possibility of contamination by inorganic anions and therefore the oxide should be an acceptable iron-making material. Thus the combination of exchange-extraction and hydrolytic stripping makes, in principle, an attractive alternative to the current aqueous phase hydrolysis processes for the removal of iron from zinc liquors. A conceptual flowsheet of the process is shown in Fig. 5.[44]

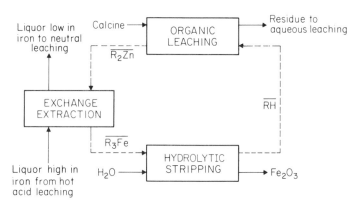

Fig. 5. Conceptual flowsheet for exchange extraction—hydrolytic stripping process.

4   Direct leaching of zinc concentrates

Processes in which the zinc sulphide concentrates are leached directly to produce zinc sulphate in solution offer distinct potential advantages over the conventional roast-leach route. Elimination of roasting to form calcine avoids the formation of zinc ferrite and therefore the attendant residue treatment problems. This leads to the possibility of treating low-grade concentrates, which are too high in iron for conventional processing, providing selective leaching of zinc is possible. More importantly however, is the fact that by proper control of the oxidation conditions in direct leaching, it is possible to convert most of the sulphide sulphur in the concentrate to elemental sulphur. Consequently the acid plant, which is necessary to convert the SO_2, produced by roasting, to sulphuric acid, is eliminated and furthermore, sulphur in elemental form offers the process operator much more flexibility than sulphur as sulphuric acid. Elemental sulphur is an inert material which can be easily transported or stored, whereas sulphuric acid has either to be used or disposed of by neutralisation.

There are a number of methods by which direct leaching of zinc sulphide can be accomplished. Leaching with ferric solutions has been the subject of a number of investigations, many of which have been reviewed by Dutrizac and Macdonald.[24] Ferric chloride solutions are particularly useful for leaching lead-containing zinc sulphide ores, because very pure lead chloride can be crystallised, merely by cooling the hot leach liquor. A recent patent assigned to Cyprus Metallurgical Processes Corporation[25] cites ferric chloride leaching, followed by lead chloride crystallisation and cementation, to produce a solution containing only zinc chloride and ferrous chloride. Zinc is removed by solvent extraction and stripped from the organic phase with a sodium chloride solution. Sodium carbonate is added to the strip liquor to precipitate zinc carbonate. The zinc-free sodium chloride is then electrolysed in a diaphragm cell to produce chlorine and sodium hydroxide. The chlorine is used to oxidise the ferrous chloride raffinate to regenerate the ferric chloride leach liquor, while the sodium hydroxide is carbonated to produce sodium carbonate used in the zinc precipitation step.

The use of sulphuric acid containing hexavalent chromium ions as a leaching agent for sulphidic zinc ores is quoted in another patent.[26] The process involves two electrolysis steps in diaphragm cells. In the first cell, the purified leach liquor, which contains only zinc and trivalent chromium, passes first into the cathode compartment where zinc is plated out and then to the anode compartment where chromium is partially oxidised back to the hexavalent state. Chromium oxidation is completed in the anode compartment of the second diaphragm cell and the regenerated liquor is then recycled to leaching.

A process similar in principle, except that manganese is used in place of chromium, is described in a paper by Fraser and Henderson.[27] Zinc sulphide concentrate is leached with return electrolyte and MnO_2 is added to oxidise the sulphide to elemental sulphur and dissolved zinc. After purification, the solution passes to a diaphragm cell, where zinc is electrowon in the cathode compartment, while manganese is oxidised and precipitated as MnO_2 in the anode compartment.

As far as is known, none of the direct leaching processes outlined above is in, or even near, commercial operation. However, one method of direct leaching of zinc sulphide concentrates is at a very advanced state of development. This is the acid pressure leaching process, developed by Sherritt Gordon Mines Ltd which is discussed in the next section.

4.1   Acid pressure leaching

Sherritt Gordon Mines Ltd. in Canada have been developing acid pressure leaching for zinc sulphide concentrates for more than twenty years. In 1959, Forward and Veltman[28] published a paper showing that the process was technically possible. The basic reaction involved is:

$$ZnS + \tfrac{1}{2}O_2 + H_2SO_4 \rightarrow ZnSO_4 + S^0 + H_2O \tag{11}$$

In order to obtain high zinc extractions, it was found necessary to maintain the temperature of reaction below the melting point of sulphur, otherwise molten sulphur coated the unreacted sulphide particles and the reaction virtually ceased. Because of the temperature limitation, retention times of several hours were necessary to achieve high zinc recoveries.

Almost a decade later, details of a modified process were published,[29] where it was shown that retention times could be cut to less than 2 hours by leaching at 150°C in the presence of excess concentrate. However this necessitated a flotation step to separate sulphur and unreacted concentrate from the leach residue, followed by sulphur distillation or hot filtration to remove sulphur from the unreacted concentrate, which was then recycled to leaching. No commercial developments resulted from this work, but Sherritt Gordon continued research on the process and in 1975 a patent[30] was issued which revealed that the addition of certain surface-active agents to the leach liquor prevented coating and occlusion of unreacted sulphide particles by molten sulphur. Additives cited as suitable include lignins, lignosulphonates, alkylaryl sulphonates and tanin compounds, particularly quebracho and other tree bark extracts. This discovery has enabled further improvements in the original process to be made, which are described in a recent paper[31] and in which it is stated that the process has been extensively piloted and is close to commercial realisation.

An outline of the process is shown in Fig. 6. In order to obtain high zinc extractions, it is necessary to grind the concentrate to 95% $< 44 \mu$m. Single-stage, continuous pressure leaching is carried out in two, three-compartment, autoclaves operated in series. The leaching solution is spent electrolyte returned from electrowinning with make-up acid to replace sulphate lost from the process in the leach residues. Typical operating conditions and results of single-stage leaching are shown in Table 3.

After leaching, the slurry is cooled by two-stage flashing to atmospheric pressure. After a solid-liquid separation and residue washing step, the liquor is passed to neutralisation and the residues to sulphur removal. Iron is removed from the leach liquor by a two-stage neutralisation. In the first stage, the liquor at 80–90°C is neutralised to pH 3·5–4 with limestone. Air is sparged into the agitated neutralisation reactor to oxidise iron, which hydrolyses and precipitates, the concentration

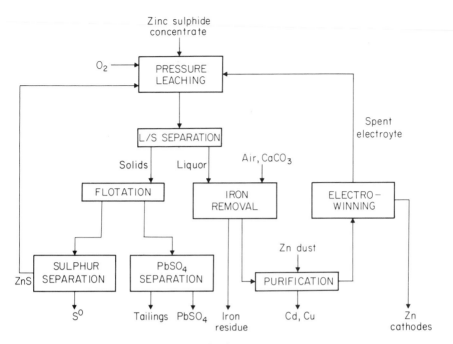

Fig. 6. The Sherritt–Gordon zinc acid pressure leach process.

Table 3. Typical leaching conditions and results[31]

Conditions:
  Max. temp. 154°C. Retention time 2 hr.
  Min. O_2 part. pressure 207 kPA (30 psi)
  Pulp density $\sim 20\%$ w/v
  Spent Electrolyte Leach Solution: Zn, 59; H_2SO_4, 183; Mg, 12 g/l.
  H_2SO_4 : ZnS mole ratio = 1·1
  Additives: Quebracho 0·2 g/l
         Calcium lignosulphonate 0·1 g/l.
Results:
  Extractions: Zn 97%, Cd 98%, Cu 68%, Fe 12%
  Conversion of S^{2-} to $S°$: 75–80%
  Conversion of S^{2-} to SO_4^{2-} : 5–10%
  Pregnant leach liquor: Zn, 170; H_2SO_4, 15; Fe, 5 g/l.

---

dropping to 0·1–0·5 g/l. Losses of Zn and Cu in the iron residue are $< 0·5\%$ and 2–10%, respectively. Removal of As, Sb, Ge and some F occurs at this stage. Retention time in the first stage is 1–3 hours. In the second-stage of neutralisation, the iron concentration is lowered to $< 0·01$ g/l by neutralisation to pH 4·5, using a 30% stoichiometric excess of ZnO, added in the form of zinc dross. Copper losses in the second stage residue are of the order of 50% and so this is recycled. Conventional purification of the iron-free liquor with zinc dust then follows, prior to electrowinning.

Sulphur is removed from the leach residues by screening and/or flotation. Subsequent treatment depends on the composition of the sulphur flotation concentrate. Melting and pressure filtration or extraction with toluene are methods suggested for recovery of sulphur, but no details are given.

Optionally, two-stage countercurrent leaching, instead of single-stage, can be used. This is the subject of a recent patent[32] and is said to give improved zinc and copper recoveries and lower free acid and iron concentrations in the pregnant leach liquor.

The capital cost of a direct pressure leach-electrowinning plant is claimed to be about 30% less than a conventional roast-leach-electrowinning plant, while the operating costs and energy requirements of the two processes are said to be similar. The reduced capital cost is almost entirely attributable to the elimination of the roasting and acid plants required in the conventional process.

The further development of the direct pressure leaching process for the treatment of complex zinc-lead sulphide concentrates, which are not treatable by conventional methods, is described in a recent paper.[33] Materials containing approximately 30% Zn; 18–22% Fe; 4–12% Pb and 0·7–0·8% Cu have been treated by both single and two-stage pressure leaching under conditions similar to those used for high-grade zinc concentrate leaching. Zinc extractions in excess of 96% are obtained. Iron reports in the leach residues, either as jarosites, or as unreacted pyrite, which is occluded in the elemental sulphur pellets formed when the leach slurry is flashed to atmospheric pressure. Lead together with any silver in the feed, also reports in the leach residue, as a mixture of lead sulphate and lead jarosite. Screening or hydrocycloning of the leach residues separates the pyrite/sulphur fraction from the lead/silver/iron fraction. The former is said to be suitable for indefinite storage, whereas the latter can be upgraded by flotation and sulphuric acid leaching to produce a lead/silver sulphate product.

Direct acid pressure leaching is undoubtedly an attractive alternative to the conventional route for zinc sulphide processing. The considerably reduced capital cost, together with the flexibility afforded by the production of sulphur in elemental form are powerful incentives for its adoption on a commercial scale. Nevertheless there are at least two aspects of the process which pose some problems. The first is the behaviour of impurities and the second is the disposal of iron.

In the conventional process, certain detrimental impurities are removed during calcining of the zinc concentrates. In particular, these include selenium, mercury, fluoride and chloride. In the direct leach process, the first three of these are said to be removed either in the leach residues or during the iron precipitation steps. It is known, however, that selenium is difficult to remove from leach liquors produced by leaching copper concentrates.[34] As selenium is very detrimental in the zinc electrowinning step, separate purification procedures may well be necessary to keep its concentration at a tolerable level in the zinc electrolyte. Any chloride in the zinc feed will report in the leach liquor and this has to be removed prior to electrowinning by either ion-exchange or silver chloride precipitation.

Iron in the feed concentrate is rejected either as unreacted pyrite mixed with elemental sulphur or as jarosites in the leach residue. The pyrite/sulphur mixture is stated to be suitable for indefinite storage, but the well-known environmental effects caused by pyrite weathering are likely to make storage of this material a less than straightforward problem. Similarly, the problems already discussed in the disposal of jarosites produced in the conventional electrolytic process will apply equally to the disposal of the leach residues from the pressure leaching process.

5   Treatment of other zinc-bearing materials

5.1   Zinc silicate ores

Two zinc silicate minerals are known – willemite Zn_2SiO_4, and hemimorphite, $Zn_4(Si_2O_7)(OH)_2.H_2O$. Smithsonite, $ZnCO_3$, is often found in association with these minerals in oxidised zinc ores, and such mixtures are often called 'calamine'. At one time, calamine ores were the major raw materials for the pyrometallurgical production of zinc, but their use declined with the growth of the electrolytic process. The reason for this is that when zinc silicates are dissolved in sulphuric acid, the solutions formed are super-saturated with respect to dissolved silica. Unless the solution conditions are carefully controlled, gelling of the silica occurs, making filtration very difficult or even impossible and leading to excessive losses of zinc-bearing solution in the residues. Thus, until recently, zinc silicate ores have been largely ignored by the electrolytic zinc industry. Evidence of revived interest, however, is provided by the fact that two major zinc companies have recently taken patents on hydrometallurgical processes for the treatment of zinc silicate ores.[35,36] Both these processes are described, together with earlier work on the processing of zinc silicate ores, in a review by Matthew and Elsner.[37]

Both willemite and hemimorphite are readily soluble in sulphuric acid to form zinc sulphate and silicic acid, e.g.:

$$Zn_2SiO_4 + 4H^+ = 2Zn^{2+} + Si(OH)_4 \tag{12}$$

Monosilicic acid condenses, by proton elimination, to form polysilicic acids, containing siloxane groups, Si-O-Si. As condensation continues, the polymers can grow to colloidal size and the colloidal particles may or may not be stable in solution. Unstable colloidal silica will further aggregate. Aggregation behaviour can range from the formation of an infinite, open-network, structure – a gel, to the formation of finite, close-packed, clusters of colloidal particles – a fine precipitate. In the former case, solid/liquid separation is virtually impossible, whereas in the latter, silica removal is possible by filtration to remove the precipitate particles. Factors which affect the stability and aggregation behaviour of solutions of colloidal silica include pH, temperature, silica concentration, impurities and ionic strength, and the key to success in the hydrometallurgical treatment of zinc silicate ores lies in the manipulation of these variables so that precipitation rather than gelation of silica occurs. Once this has been achieved, silica can be removed by filtration and the resulting zinc sulphate solution can then be purified and electrowon in the normal way.

In the process devised by Electrolytic Zinc of Australasia (E.Z.), rapid neutralisation of the acidic leach liquor is used to precipitate silica in an easily filtered form. The process has been extensively tested on Australian willemite ores[38] and Thai hemimorphite-smithsonite ores[39] and it is reported that the process will be used in an electrolytic zinc plant, with a capacity of 60 000 tons Zn per annum, to be built in Thailand.[39] A simplified flowsheet of the Thai plant is shown in Fig. 7.

The ore, averaging 25% Zn, is wet ground with neutral solution at 55% solids. Leaching is carried out continuously in three agitated tanks with return electrolyte,

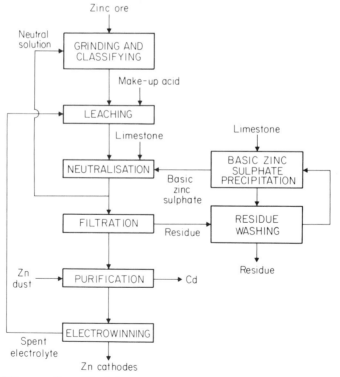

Fig. 7. The EZ process for oxidised zinc ores.

averaging 150 g/l H_2SO_4 and 50 g/l Zn, and make-up acid. A retention time of $4\frac{1}{2}$ hours at 40 – 50°C is required and acid additions are controlled so that the final leach liquor has a pH of about 2. The liquor, which contains 15 g/l SiO_2, is sufficiently acidic to prevent polymerisation of the silica. The next, and key, step is neutralisation. This is done continuously in a series of three agitated tanks. The leach pulp flows into the first of these, which is held at 60 – 65°C and a pH of about 4·3 by addition of limestone and/or basic zinc sulphate. Under these conditions, silica, together with other impurities such as iron and arsenic, is rapidly precipitated and gel-formation is avoided. Retention time in the neutralisation section is the same as in leaching. Filtration of the pulp produces a clear neutral solution which goes forward to standard purification with zinc dust and then to electrolysis. The residue from filtration is washed to remove soluble zinc and the wash solution is treated with limestone to produce the basic zinc sulphate used in neutralisation. Overall zinc recovery for the Thai ore is estimated at 90%.

The approach taken by Vieille-Montagne to zinc silicate processing differs in that silica precipitation is made to occur during the leaching operation.[35] This is done preferably in a continuous leaching system, consisting of at least four stages. Return electrolyte, containing between 100 and 200 g/l H_2SO_4, is added progressively to each of the stages and the acidity increases from stage to stage, with the final liquor containing from 1·5 to 15 g/l free acid. Retention time is at least three hours and the temperature is maintained between 70 and 90°C. After leaching, the pulp is agitated at the same temperature for a further two to four hours to complete the precipitation of silica. Final silica concentrations in the range 0·1 to 0·3 g/l result. Prior to purification with zinc dust, the leach liquor is neutralised to a pH of between 3 and 5 over a period of 1 to 3 hours by addition of lime or zinc oxide. Air is sparged in during neutralisation to oxidise and precipitate iron and other impurities.

A rather different approach is the so-called 'quick-leach' method described in a paper by Dufresne.[40] This involves the addition of carefully controlled amounts of water and sulphuric acid to the siliceous zinc ore so that water is completely consumed by the formation of hydrated zinc sulphate. During the reaction, silicic acid is dehydrated and reports in the solid as granular silica. The principle of the process is illustrated by the following equation:

$$Zn_2SiO_4 + 2H_2SO_4 + 12H_2O \rightarrow 2ZnSO_4.7H_2O + SiO_2 \qquad (13)$$

This reaction takes place very rapidly. The slurry of ore, acid and water is said to set and form a crumbly solid within about one minute of mixing. This can then be leached with water to dissolve the metal sulphates and the silica remains in the residue in an easily filterable form. Small-scale experiments showed that return electrolyte from electrowinning could be used in place of water to leach the sulphated ore. This indicated that the process could be integrated into existing electrolytic plants to treat the oxidised zinc ores which are often found in association with zinc sulphide deposits.

The integration of silicate ore processing into existing sulphide-based electrolytic zinc plants is an attractive method of increasing metal production because it avoids having to increase roaster and acid plant capacity, which usually accounts for 25–30% of the capital cost of an electrolytic plant. Furthermore silicate leaching offers an outlet for sulphuric acid produced in the acid plant, although sulphate would have to be removed from the circuit at some point to maintain the sulphate balance.

5.2   Pyrites Cinders

A new process of considerable technical interest is the Espindesa process for the extraction of zinc from pyrites cinders. This process utilises solvent extraction to recover zinc from leach liquors arising from the chloride roasting of pyrites cinders and it is the first commercial application of a solvent extraction-electrowinning route for the primary production of zinc. The use of solvent extraction in the hydrometallurgy of zinc has been hindered by the stringent impurity limits that must be met to produce an electrolyte suitable for electrowinning. The Espindesa process achieves the required high purity in the electrolyte by utilising two cycles of solvent extraction. The first cycle involves extraction of zinc as an anionic chlorocomplex by an amine extractant. This is followed by extraction of cationic zinc from the amine strip liquor by an alkyl phosphoric acid. In this way, zinc is separated from all deleterious impurities and can be electrowon directly from the liquor produced by stripping the alkyl phosphoric acid.

The process was developed in Spain by Espindesa and Tecnicas Reunidas S.A. and a plant using this process, with a capacity of 8000 tonnes/year of electrolytic zinc was built at Bilbao and came on stream in August 1976. There is little doubt that the successful commercial operation of this process will be seen as a landmark in the development of zinc hydrometallurgy and that solvent extraction will play an increasing part in zinc processing, particularly where complex or low-grade zinc-bearing materials are involved.

A flowsheet of the Espindesa process is shown in Fig. 8. A full description of the process has not yet been published and the information that follows has been taken from publicity bulletins.[41,42] The process technology is covered by a number of Spanish patents.[43,45]

The feed to the plant is a liquor resulting from the leaching of pyrites cinders after chloride roasting. The feed liquor contains 25–30 g/l Zn, 20–25 g/l Fe, 70–90 g/l chloride, 100–120 g/l sulphate plus quantities of Cu, Cd, Co, Pb, Ni, As, Mn, other less significant metals, and alkali and alkaline-earth metals. Ferric iron

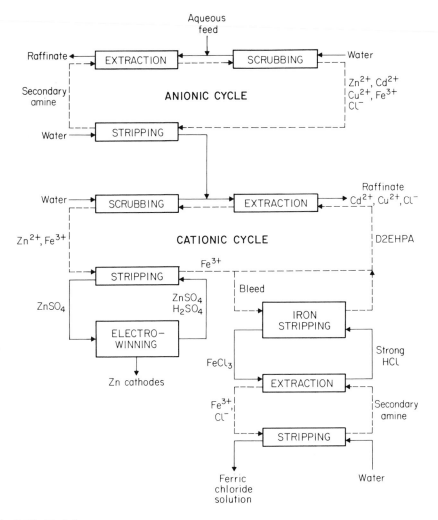

Fig. 8. The Espindesa process.

follows zinc through the process and therefore any iron in the feed must be reduced to the ferrous state prior to solvent extraction.

Zinc is first extracted as a chlorocomplex in the anionic solvent extraction cycle, using a secondary amine extractant. A centre-feed system is used, with scrubbing with acidified water to remove entrained feed liquor and metal ions co-extracted with the zinc. Zinc is stripped from the loaded amine with water to produce a zinc chloride solution. Metals accompanying zinc at this stage are principally Cu^{2+}, Cd^{2+} and Fe^{3+}. The aqueous raffinate from amine extraction typically contains 0·1 g/l Zn and 10 ppm entrained organic phase after suitable coalescence.

Zinc is then re-extracted from the zinc chloride strip liquor in the cation extraction cycle using a solution of di-2-ethylhexyl phosphoric acid (D2EHPA) in kerosene. The pH in the extraction section is carefully controlled at 2·60 by introduction of ammonia. A centre-feed system is again used with scrubbing of the loaded organic phase to remove entrained aqueous phase and, in particular, chloride ions. The raffinate from the cationic extraction cycle contains all the copper and cadmium extracted in the first cycle, whereas any ferric ion will accompany zinc into the organic phase. Zinc is stripped from the loaded D2EHPA by contact with return electrolyte from electrowinning. The strip liquor is the pregnant electrolyte feed for electrowinning and contains 50 – 60 g/l Zn, 20 ppm Fe, 30 ppm Cl^- and < 1 ppm of metallic impurities such as Cu, Cd, Co, As, etc. Organic entrainment in the electrolyte can be controlled at 1 – 5 ppm using settling tanks and charcoal filters. Ferric iron in the organic is not stripped with the zinc and so a continuous bleed of organic from the D2EHPA circuit is required to prevent build-up of iron in the organic phase. Iron is removed from the D2EHPA by stripping with concentrated hydrochloric acid. In order to conserve the concentrated acid for recycle to stripping, iron is re-extracted, using a secondary amine, which is presumably then stripped with water to give a solution of ferric chloride.

A recent improvement of the process, which has been pilot-tested, is the use of milk of lime, instead of ammonia, as a neutralising agent for pH control during the cationic extraction cycle. This is said to offer a significant reduction in production costs.

The two-cycle solvent extraction technique used in the Espindesa process provides considerable flexibility with regard to feed liquor composition. The major requirements are firstly that sufficient chloride ion must be present in the feed liquor to cause extraction of zinc by the amine and secondly the concentration of ferric ions must be low. The first condition can be met, if necessary, by addition of chloride in any readily available form and the second can be met by prior reduction of iron in solution to the ferrous state. Thus the process is capable of development to treat a variety of feedstocks. In particular, the treatment of solutions arising from direct leaching of zinc sulphide concentrates with, for example, ferric chloride solutions is an obvious potential application of this process, which may eventually result in an economic alternative to the conventional roast-leach-electrowin route.

6   References

1 Lewis H.M. The outlook for zinc. *CIM Bull.*, 1978, **71** (789), 130–133.
2 Meisel G.M. New generation zinc plants, design features and effect on costs. *J. Metals*, 1974, **26** (8), 25–32.
3 Asturiana de Zinc S.A. Spanish Patent 304 601, Appln. Oct. 12, 1964.
4 Electrolytic Zinc Company of Australasia Ltd., Australian Patent 401 724, Appln. Mar. 31, 1965.
5 Det Norske Zinkkompani A/S, Norwegian Patent 108 047, Appln. Apr. 30, 1965.

6 Gordon A.R. Improved use of raw material, human and energy resources in the extraction of zinc. In *Advances in Extractive Metallurgy*. London: Institution of Mining and Metallurgy, 1977, 153–160.

7 Steintveit G. Electrolytic zinc plant and residue recovery Det Norske Zinkkompani A/S. In World Symposium on the Mining and Metallurgy of Lead and Zinc. New York: AIME, 1970, 223–246.

8 Haigh C.J. and Pickering R.W. The treatment of zinc plant residues at the Risdon Works of the Electrolytic Zinc Company of Australasia Ltd., ibid, 423–447.

9 Steintveit G. Treatment of zinc leach plant residues by jarosite process. In *Advances in Extractive Metallurgy and Refining*. London: Institution of Mining and Metallurgy, 1971.

10 Wood J. and Haigh C. Jarosite process boosts zinc recovery in electrolytic plants. *World Min.*, Sept. 1972, **25**, 34–38.

11 Gordon A.R. and Pickering R.W. Improved leaching technologies in the electrolytic zinc industry. *Metall. Trans.*, 1975, **6B**, 43–53.

12 Outokumpu Oy. British Patent 1 464 447, Appln. Feb. 12, 1973.

13 Huggare T-L, Fugleberg S. and Rastas J. How Outokumpu Conversion Process raises Zinc recovery. *World Min.*, Feb. 1974, **27**, 36–42.

14 Societe de la Vieille Montagne. Belgian Patent 724 214, Appln. Nov. 20, 1968.

15 Van Den Neste E. Metallurgie Hoboken-Overpelt's zinc electrowinning plant. *CIM Bull.*, 1977, **70** (784), 173–185.

16 Electrolytic Zinc Company of Australasia Ltd. Australian Patent 424 095, Appln. May 15, 1970.

17 Tsunoda S., Maeshiro I., Emi F. and Sekine K. The construction and operation of the Lijima electrolytic zinc plant. Paper presented to AIME Annual Meeting, Chicago, 1973. TMS Pap. A73–65.

18 Anon. Environmental considerations and the modern electrolytic zinc refinery. *Mining Engng.* Nov. 1977, 31–33.

19 Steintveit G. and Lindstad T. Smelting of lead-silver residue and jarosite precipitate. In *Advances in Extractive Metallurgy*. London: Institution of Mining and Metallurgy, 1977, 7–12.

20 Thorsen G. U.S. Patent 4 008 134, 1977, British Patent 1 474 944, 1977.

21 Shell. German Patent 2 404 185, Jan 29, 1974.

22 Van der Zeeuw A.J. Purification of zinc calcine leach solutions by exchange extraction with the zinc salt of Versatic acid. *Hydrometallurgy*, 1976/7, **2**, 275–284.

23 Thorsen G. and Monhemius A.J. Precipitation of metal oxides from loaded carboxylic acid extractants by hydrolytic stripping. Paper presented to AIME Annual Meeting, New Orleans, 1979. TMS Pap. A79–12.

24 Dutrizac J.E. and MacDonald W.A. Ferric iron as a leaching medium. *Minerals Sci. Engng.*, 1974, **6** (2), 59–100.

25 Goens D.N. and Reynolds J.E. U.S. Patent 3 973 949, 1976.

26 Morris T.M. and Bilson E.A. U.S. Patent 3 923 617, 1975.

27 Fraser D.B. and Henderson J. MnO_2 in hydrometallurgy – electrowinning of zinc from sulphides. In Metals 77, 30th Ann. Conf. Aust. Inst. Metals, 1977, 9B4–9B5.

28 Forward F.A. and Veltman H. Direct leaching zinc sulphide concentrates by Sherritt Gordon. *J. Metals*, 1959, **11**, 836–840.

29 Veltman H. and O'Kane P.T. Accelerated pressure leaching of zinc sulphide concentrates. Paper presented to AIME Annual Meeting, New York, 1968.

30 Kawulka P., Haffenden W.J. and Mackiw V.N. Recovery of zinc from zinc sulphides by direct pressure leaching. U.S. Patent 3 867 268, Feb. 18, 1975.

31 Doyle B.N., Masters I.M., Webster I.C. and Veltman H. Acid pressure leaching of zinc concentrates with elemental sulphur as a by-product. Paper presented to XI Commonwealth Mining. Met. Cong., Hong Kong, 1978.

32 Veltman H., Mould G.J.J. and Kawulka P. Two-stage pressure leaching process for zinc and iron-bearing mineral sulphides. U.S. Patent 4 004 991, Jan. 25, 1977.

33 Bolton G.L., Zubrycky J.N. and Veltman H. Pressure leaching process for complex zinc-lead concentrates. Paper presented to XIII Int. Mineral Proc. Cong., Warsaw, 1979.
34 Swinkels G.M. and Berezowsky R.M.G.S. The Sherritt–Cominco Copper process – Part 1: The process. *CIM Bull.*, 1978, **71** (790), 105–121.
35 Societe de la Vieille Montagne. Belgian Patent 803 156, Appln. Aug. 2, 1973.
36 Electrolytic Zinc Company of Australasia Ltd. Australian Patent 407 500, Appln. Mar. 13, 1968.
37 Matthew I.G. and Elsner D. The Processing of zinc silicate ores – a review. *Metall. Trans.*, 1977, **8B**, 85–91.
38 Matthew I.G. and Elsner D. The hydrometallurgical treatment of zinc silicate ores. *Metall. Trans.*, 1977, **8B**, 73–83.
39 Wood J.T., Kern P.L. and Ashdown N.C. Electrolytic recovery of zinc from oxides and ores. *J. Metals*, 1977, **29** (11), 7–11.
40 Dufresne R.E. Quick leach of siliceous zinc ores. *J. Metals*, 1976, **28** (2), 8–12.
41 Tecnicas Reunidas S.A. Zinc recovery process. Bulletin available from Non-Ferrous Metallurgical Divn., Arapiles 13, Madrid – 13.
42 Davy Powergas Ltd. Espindesa zinc solvent extraction process. Bulletin available from Non-Ferrous Metallurgical Divn., 49 Wigmore Street, London W1H 9LE.
43 Diaz Nogueira E. *et al.* Spanish Patents: 403 506, Apr. 16, 1975; 405 759, Oct. 16, 1975; 441 536, Apr. 1, 1977; 458 700, Mar. 1, 1978.

References added in proof:

44 Thorsen G. and Grislingas A. Solvent extraction of iron in zinc hydrometallurgy. Paper presented to AIME Annual Meeting, Las Vegas, 1980 (This paper discusses the integration of solvent extraction of iron into the electrolytic zinc process).
45 Nogueira E.D., Regife J.M. and Blythe P.M. Zincex – the development of a secondary zinc process. *Chem. Ind. (London)*, 1980 (2), 63–67. (This paper deals with the Espindesa process).

Index